Understanding China

The series will provide you with in-depth information on China's social, cultural and economic aspects. It covers a broad variety of topics, from economics and history to law, philosophy, cultural geography and regional politics, and offers a wealth of materials for researchers, doctoral students, and experienced practitioners.

Lijin Yan

Value-Based Civilization

A New Era of Human Civilization

中国工信出版集团
電子工業出版社
PUBLISHING HOUSE OF ELECTRONICS INDUSTRY
http://www.phei.com.cn
Springer

Lijin Yan
Beijing, China

ISSN 2196-3134 ISSN 2196-3142 (electronic)
Understanding China
ISBN 978-981-96-9489-1 ISBN 978-981-96-9490-7 (eBook)
https://doi.org/10.1007/978-981-96-9490-7

Jointly published with Publishing House of Electronics Industry
The printed edition is not for sale in China Mainland. Customers from China Mainland please order the print book from: Publishing House of Electronics Industry.
ISBN of the Co-Publisher's edition: 978-712-14-7291-6

This Springer imprint is published by the registered company Springer Nature Singapore Pte Ltd.
The registered company address is: 152 Beach Road, #21-01/04 Gateway East, Singapore 189721, Singapore

Foreword by Lu Jun

Value-Based Civilization: Leading Humanity into a New Epoch

In today's era, the wave of digital revolution is sweeping across the globe with overwhelming momentum, profoundly reshaping every aspect of human society. A profound and far-reaching transformation has already begun, and humanity stands at the threshold of a new era. Reflecting on history, from the nomadic civilization to agrarian civilization and then to industrial civilization, each shift has propelled tremendous leaps in productivity and major transitions in human society, creating brilliant chapters in the annals of human development. Now, amidst the burgeoning wave of information civilization, the rapid development of digital technology is injecting both powerful momentum and numerous uncertainties into this process, accelerating humanity's journey into uncharted territories.

At this pivotal historical juncture, we are prompted to reflect deeply: What is happening to the world? Where should humanity go? The widespread application of digital technologies has brought about profound changes in our lives. From convenient online shopping and efficient remote work to abundant online educational resources and precise intelligent medical diagnostics, digital technology has permeated every corner of life, transforming our lifestyles, work patterns, and modes of social interaction. It has also, imperceptibly, influenced our ways of thinking and values. However, this process is not without challenges. Issues such as information security, privacy breaches, and online fraud have emerged, while the digital divide exacerbates social inequality. In such a complex and ever-changing era, characterized by both opportunities and challenges, we urgently need a new philosophy and wisdom to guide our direction and illuminate humanity's path forward.

My friend, Dr. Lijin Yan, with his keen insight and profound thought, has proposed the pioneering concept of "Value-Based Civilization," providing a key to answering the questions of our times and opening the door to exploring the future possibilities of the world.

Dr. Lijin Yan, a distinguished scholar and entrepreneur with outstanding achievements in both academia and business, has long focused on the dynamic development of the digital technology revolution and its profound impact on the world. Leveraging his solid academic foundation and rich practical experience, he delves into the logic behind digital technologies, reflecting on how to advance the progress and prosperity of world civilization in this era. His book, *Value-Based Civilization: A New Era of Human Civilization*, offers unique and profound insights, providing a fresh perspective to understand the world and our mission within it.

In this significant work, Dr. Yan thoroughly analyzes the essence of the digital technology revolution and its transformative mechanisms for the world. He points out that digital technology not only alters economic operating models, spawning numerous emerging industries such as e-commerce, digital finance, and the sharing economy, but also has far-reaching implications for social governance, cultural dissemination, education, and healthcare. However, technological development should not merely focus on superficial innovation and progress but must also emphasize the underlying values and humanistic concerns. Thus, he creatively proposes the new civilizational paradigm of "Value-Based Civilization."

"Value-Based Civilization" places the comprehensive development of humanity at its core, emphasizing fairness and justice as its foundation while pursuing human dignity, happiness, and freedom. Within the framework of Value-Based Civilization, the application of technology must be human-centered, considering its impact on people and ensuring that technological advancements benefit all humanity rather than exacerbating social inequalities. For instance, in promoting digital technology, attention must be paid to the digital inclusion of disadvantaged groups and their training in digital skills to bridge the digital divide and allow everyone to enjoy the dividends brought by technological progress.

Simultaneously, Dr. Yan calls for an open and inclusive mindset to transcend national and cultural boundaries. In the digital age, the world is becoming more interconnected, and exchanges and integration between different nations and cultures are increasingly frequent. We should respect cultural diversity, learn from one another, and work together to address global challenges. He firmly believes that only through global cooperation and collective effort can the potential of digital technology be fully realized, creating a better future for all.

From the long history of human civilization, shared development has always been the only path to achieving global value consensus. Amid the tide of digital technology, the interdependence and interconnectedness between nations and regions have become increasingly significant. No country can navigate the digital technological transformation in isolation. Both developed and developing countries possess their unique strengths and challenges in the digital domain. Only through strengthened global cooperation, resource sharing, and mutual complementarity can we collectively tackle the challenges of the digital age and propel human society towards greater prosperity.

Looking back at history, every major transformation of human civilization has been accompanied by shifts in commercial, trade, and financial operations. Today, the emergence of Value-Based Civilization is likewise accelerating changes in every aspect of human life. Although the advancement of digital technology has led to leaps in productivity, it has also triggered conflicts and changes in production relations within the international community. For example, during past technological revolutions, the agrarian civilization era witnessed nations competing for land resources, while the industrial revolution gave rise to colonial expansion and imperialist wars. Now, the rapid development of digital technology has led to the obsolescence of traditional industries and changes in employment structures, posing challenges to social stability. Furthermore, digital technology is disrupting traditional international rules and orders, with varying speeds of technological development across nations intensifying international competition and cooperation. How to reconcile differing interests and establish international rules and orders suitable for the digital era has become a major task for humanity.

It is noteworthy that quantum technology is emerging as a new variable and opportunity for the future of human civilization. Quantum technology shows immense potential in fields such as computing, communication, and encryption, promising exponential improvements in computational power, absolute security in communication, and deeper understanding of the microscopic structure of matter. However, its development also brings new challenges, such as potential threats to existing encryption systems, the need to establish global standards and safety mechanisms for quantum communication, and concerns about its militarization potentially triggering a new arms race. These issues extend beyond technology, touching upon global security, ethics, and the restructuring of international rules.

Within the framework of Value-Based Civilization, we must guide the development of quantum technology with a prudent and responsible attitude. On the one hand, we must harness quantum technology's positive contributions to scientific research, computational efficiency, and communication security, creating more value for humanity. For instance, leveraging quantum computing to accelerate drug discovery, solve complex scientific problems, and advance artificial intelligence; or utilizing quantum communication to safeguard critical information infrastructure, providing reliable communication for finance, governance, and other sectors. On the other hand, we must be vigilant about the negative impacts of quantum technology, fostering international cooperation and coordination to jointly establish ethical guidelines, safety standards, and international rules. Nations must abandon unilateralism and zero-sum thinking, instead pursuing multilateral consultations and cooperative mechanisms to ensure that the development of quantum technology aligns with the shared interests of humanity, preventing global crises arising from its misuse.

As a friend of Dr. Lijin Yan, I am deeply honored to write the foreword for his book. I firmly believe that Value-Based Civilization: A New Era of Human Civilization will provide a profound framework for thought, helping us better understand the world and our mission, and contributing wisdom and strength to global prosperity.

Here, I express my respect for Dr. Lijin Yan's dedicated efforts and unique insights, hoping this book will gain widespread attention and provoke in-depth reflection worldwide, making a positive contribution to the development of world civilization. Let us join hands, guided by Value-Based Civilization, and courageously embrace the tides of the digital and quantum age. Together, we can strive to reach new heights of human civilization and create a more prosperous, harmonious, and beautiful future.

Beijing, China

Lu Jun
Academician, Chinese Academy of Engineering

Foreword by Wang Zhongming

Harmony and Civilization: Ushering in a New Chapter of Human Development with Exceptional Wisdom

Throughout the long journey of human civilization, every major technological breakthrough has left a distinctive mark of its era. From the invention of the printing press that heralded a new age of cultural dissemination to the industrial revolution that reshaped production and societal structures, the current wave of digital technological revolution is profoundly transforming the world. It is elevating human civilization to new heights while placing us at the crossroads of opportunities and challenges.

Amidst the digital surge, life as we know it is being comprehensively redefined. While marveling at technological progress, we cannot help but feel uncertain about the future's direction. It is in this context that Value-Based Civilization stands out as a remarkable work, offering both practical significance and intellectual value in unraveling the questions of our time and illuminating the way forward.

The author of this book, Dr. Lijin Yan, is an exceptional figure with achievements spanning academia, commerce, and societal insight. He has previously authored several well-received works, including From Harmonious China to a Harmonious World, Harmonious Development: Thirty Years of Chinese Agriculture, and Great Power Harmony. His profound understanding of the concept of "harmony" is evident in his writings. Harmony and civilization, after all, are deeply intertwined. Any genuine civilization must aspire to a specific form of harmonious perfection, with harmony serving as its value orientation. From this perspective, Value-Based Civilization represents a new exploration and leap forward, building upon his previous intellectual achievements.

With a solid foundation, rich experience, and precise grasp of contemporary trends, Dr. Yan delves into the core of the digital technological revolution. After years of dedicated research, he has distilled the concept of "Value-Based Civilization," showcasing a remarkable capacity for philosophical and strategic thinking.

Today, digital technology weaves an invisible web that permeates every corner of life. From the everyday use of smart devices and the proliferation of remote work and online education to digital financial transactions and innovative medical applications, its influence is omnipresent, reshaping traditional behaviors and thought patterns. However, the rapid advancement of technology has also given rise to numerous challenges, including privacy breaches, data security risks, doubts about information authenticity, ethical dilemmas, and even the specter of technological cold wars.

Dr. Yan insightfully observes that we must not be blinded by the superficial allure of technology but instead explore the value foundations and civilizational trajectory underpinning it. The concept of "Value-Based Civilization" breaks through the limitations of traditional civilizational studies by linking digital technology to human values. It emphasizes that technological progress must return to addressing fundamental human needs and spiritual aspirations. This means that while enjoying the convenience digital technology brings to life, we must also focus on shaping and perpetuating core values such as fairness, justice, integrity, and inclusivity. For instance, in the digital economy, ensuring fairness in e-commerce transactions, neutrality in algorithmic recommendations, and security in digital finance are all integral to the guidance of Value-Based Civilization.

Grounded in the transformative forefront of digital technology, this book fully acknowledges the extraordinary contributions of artificial intelligence in enhancing productivity and optimizing service experiences. At the same time, it provides solutions and pathways to address associated challenges, such as shifts in employment structures, concerns about human creativity, and the ethical dilemmas of responsibility. Through case studies, logical reasoning, and theoretical exposition, the author demonstrates the critical role of Value-Based Civilization in complex technological contexts. The book guides us to embrace technology while upholding humanity and moral principles, achieving harmonious coexistence between technology and humanity.

With a broad global perspective, the book cautions against the exacerbation of developmental imbalances brought about by digital technology, even as it connects the world more closely. The advocacy for Value-Based Civilization underscores the importance of constructing a shared value system that transcends national, cultural, and class boundaries, promoting global harmony, coexistence, and coordinated development. Regardless of their level of development, all countries can find a foundation for collaboration and a direction for progress within this framework. Together, they can tackle global challenges such as climate change, public health crises, and cybersecurity threats. Upholding the principles of Value-Based Civilization, nations can strengthen cooperation in technology development, resource sharing, and exchange of expertise, ensuring steady progress for human civilization in the digital era.

Value-Based Civilization is not just a theoretical treatise but also a practical guide to action. With exceptional wisdom, relentless effort, and a deep concern for the destiny of humanity, Dr. Lijin Yan presents this masterpiece, a product of our times and a crystallization of profound insight. It carries the mission of advancing human civilization.

At a time when the world is undergoing accelerated transformation, works like Value-Based Civilization that explore and illuminate the path forward are more necessary than ever. May every reader draw from its wellspring of wisdom, allowing the light of Value-Based Civilization to guide the way forward and lead humanity toward a higher stage of civilization.

Beijing,China

Wang Zhongming
Member, Advisory Committee
of the All-China Federation of Industry
and Commerce Think Tank

Foreword by Ahmet Mücahid Ören

Towards a New Era of Human Value-Based Civilization

As we look back at the long journey of human civilization, it's clear that we've always sought progress and direction.

Today, we find ourselves in the midst of a sweeping digital revolution—a turning point filled with both extraordinary opportunities and significant challenges.

At this critical juncture, Mr. Lijin Yan has offered us a thought-provoking and timely concept: "Value-Based Civilization."

When we reflect on history, values have always been the heart and soul of civilization. They are the shared beliefs and aspirations that unite humanity, much like the timeless threads that hold the fabric of society together.

Meanwhile, technology has always been the driver of change, shaping how we live and interact with the world. From the industrial revolution to the digital revolution we are experiencing now, the relentless forward progress in technology has had the power to transform not only economies but also the very essence of our lives.

Yet in this new era, the role of businesses—and economies at large—must evolve.

Mr. Lijin Yan insightfully reminds us that businesses cannot remain solely focused on generating profits. They have a larger responsibility: to promote and uphold values that matter. These values extend beyond economics to encompass social responsibility, environmental care, and a deep commitment to humanity. By integrating these ideals into their vision, businesses can align economic success with social progress, creating a more sustainable and meaningful path forward.

The wave of digital innovation we are witnessing today offers unparalleled opportunities to rethink and reshape our future. But innovation is not just about new products or services. It's also about cultural enrichment, reimagining business models, and redefining what we stand for. The interconnectedness, intelligence, and data-driven insights enabled by technology have become essential tools for navigating the challenges of our time and unlocking new possibilities.

This era of change is more than a technological revolution—it's an awakening of human consciousness.

As we strive for rapid economic growth, we must also consider how to cultivate values that uplift humanity and stand the test of time. The goal is to build a fairer, freer, and more equitable global community—a shared space where human values are celebrated and strengthened. By doing so, we can face the challenges of the present while paving the way for a more harmonious and prosperous future.

Mr. Lijin Yan's concept of "Value-Based Civilization" is both intriguing and compelling. It challenges us to look beyond short-term gains and focus on the bigger picture: the enduring progress of human civilization.

At its core, civilization is about more than material advancement. It's about improving lives, building communities, and ensuring that all people can thrive—spiritually, mentally, and physically.

True progress requires a commitment to values that guide us toward this ideal. When science and technology are used to enhance lives and strengthen communities, they bring immense benefits. But if they are used to destroy, oppress, or exploit, they risk undermining the very foundations of civilization.

With these reflections, I find this book's central vision to be an intriguing and compelling exploration of how shared human values can shape the future.

In a time of globalization, it encourages us to advance the common values of humanity and to embrace the idea of a global community with a shared future.

Ankara, Türkiye

Ahmet Mücahid Ören
Chairman, İhlas Holding Inc.

Foreword by Cemil Ceyhan Aral

For Value-Based Civilization

Advancement in technology is due to addition of new ideas to the ideas of the previous generations.

A "Prefect Civilized Person", who has good wisdom and ethics and is advanced in the scientific knowledge, latest industries of his time is called a "Civil or a Progressive Civilized Person."

Alternately, a person who is advanced in scientific knowledge and has established basic industries, and yet who has poor morals, is called a "Tyrant", a "Regressive Person", a "Bandit", or a "Dictator."

Those who are hiding behind, only in the "scientific knowledge" and "civilization" terms or words but; have poor morals, ethics are called "Savages or Vulgar People."

Civilization means building cities, technology, industries, and but along with it, serving all human beings and all human rights make every society comfortable and prosper.

In short, when the scientific knowledge, technology, industry and arts are accompanied with beautiful morals, it is called 'a perfect civilization'.

A civilized person uses scientific knowledge and arts for serving humanity and prosperity.

Alternately, tyrants use them for torturing, harming human beings and societies.

We can see that a true civilized person is a progressive person. Uncivilized person is a retrogressive and miserable person.

People's living in peace and comfort is only possible through ethics of good values of perfect civilized persons.

Let it be known that good values order us to have beautiful morals, mutual love for each other, respect for the elders and compassion for younger people, and to be kind to everyone, irrespective of their ethnic groups or opinions.

Civilized and good valued people, Pay everyone's due rights and wages to their workers! Never violate the laws of the countries and do not oppose orders regulations of the governments! Pay their taxes on their due times!

We must generate the righteous and civilizes people everywhere!

I see the book of Value-Based Civilization of My best Friend, Mr. Lijin Yan, serves to open a new horizon into the minds of Young Generation coming after us.

Let us love and help each other, so that the World become a Paradise for all of us.

Ankara, Türkiye

Cemil Ceyhan Aral
Chairman, Aral In

Preface

In today's 21st century, we humans are immersed in an unprecedented digital wave. Like standing on the shore of a surging ocean of information, we deeply feel the strong impact of information technology everywhere. Information technology is like a surging torrent, sweeping across every corner of the globe with an unstoppable force. From the details of our daily lives to the layout of the macroeconomy and from the profound changes in social structures to the grand narrative of the evolution of human civilization, digital technology has left a profound and indelible mark. Digital technology has broken down the spatial and temporal boundaries of information dissemination, making it possible for people separated by oceans to communicate in real time, as if the entire world has been condensed into a "global village." We can easily video call with distant friends and family to share the details of our lives; we can follow news events from around the world on social media and stay informed about the latest developments in the world. This immediacy and boundlessness of information have greatly closed the distance between people, making the world more connected and interdependent. In the production sector, digital technology has driven innovation in production methods. Automated and intelligent production lines have significantly increased production efficiency, completely reshaping the industrial landscape. Robots and intelligent systems have replaced many traditional manual jobs in factories, not only increasing production speed but also reducing costs and improving product quality. In agriculture, the application of precision agriculture technology has made crop cultivation more scientific and efficient, allowing farmers to manage every aspect of crop growth more precisely through data analysis and intelligent monitoring. The rise of emerging consumer models such as online shopping and digital entertainment has profoundly changed people's consumption habits, becoming an indispensable part of life. We can easily purchase goods from all over the world online, enjoying a convenient shopping experience; we can watch the latest movies and TV shows at home through streaming media platforms, enjoying a rich entertainment life. These emerging consumer models not only provide consumers with more choices and convenience but also bring new business opportunities and markets for merchants. It can be said that the profound social changes triggered by digital technology have spread to every corner of the globe, touching the very foundations of society. Education, healthcare, transportation, finance, and other industries are all undergoing digital transformation, a grand and journey of civilization rebuilt. Digital technology has not only changed our way of life but is also changing our ways of thinking and behavior. In this digital age, each of us is both a participant and a witness, collectively experiencing this magnificent transformation.

Beijing, China

Lijin Yan

Competing Interests The author has no competing interests to declare that are relevant to the content of this manuscript.

Contents

Part I
The Evolution Course of Human Civilization Form

Chapter 1
The Origin of Human Civilization

In the vast, deep, and mysterious river of the universe, the Earth stands out as a brilliant blue gem, silently cradling the miraculous process of life's gestation and reproduction. Among the countless stars and dark planets scattered throughout the cosmos, it is the Earth, with its unique blend of conditions, that has become the cradle of life. From the vantage point of space, the expansive blue oceans, the continuous stretches of green land, and the ever-changing white clouds together paint a dreamlike picture, constructing the most magnificent stage for the emergence of life.

The Earth, with its perfect balance of water, air, and land, has nurtured a diverse array of life forms, each playing a crucial role in the intricate web of ecosystems. The lush forests, teeming with biodiversity, the vast savannas, and the deep oceans, all teeming with life, are testaments to the Earth's unique ability to sustain life. The delicate dance of the seasons, the nurturing warmth of the sun, and the protective embrace of the atmosphere all contribute to the symphony of life that plays out on this blue planet.

And human beings, as the stewards of this unique intelligent light on Earth, have harnessed their ingenuity to create a rich tapestry of human civilization. Through a series of magnificent and exciting social practice activities, humans have built cities that rise like beacons of progress, crafted technologies that push the boundaries of what is possible, and developed cultures that celebrate the diversity and richness of human experience.

When these elements converge, like a trickle of water that grows into a mighty river, they accumulate to a specific quantity and quality, and achieve the exciting moment of qualitative change. It is during these moments that human beings resolutely break the shackles of ignorance and superstition, forging ahead into the light of civilization and wisdom. The invention of writing, the discovery of fire, the wheel, and the steam engine are all pivotal moments in human history that signify this transformative journey.

Since these monumental breakthroughs, people have become civilized beings, and human society has marched into the era of civilization. The creation of agriculture led

L. Yan, *Value-Based Civilization*, Understanding China,
https://doi.org/10.1007/978-981-96-9490-7_1

to the establishment of permanent settlements, the rise of cities, and the development of complex societies. The invention of the printing press democratized knowledge, while the industrial revolution reshaped the world's economy and society, leading to unprecedented technological advancements.

A magnificent epic about the origin of civilization has opened the grand prelude. This epic is not just a tale of human triumph over the forces of nature, but also a story of continuous learning, adaptation, and growth. It is a narrative of how humans have learned to harness the power of the natural world, to build communities, to create art, and to explore the mysteries of the universe. It is a story of resilience and innovation, of the indomitable human spirit that continues to push the boundaries of what is possible, and to dream of a future where life not only persists but thrives.

1.1 The Cornerstone of the Origin of Human Civilization: Natural Environment and Biological Evolution

1.1.1 The Foundation of Nature: The Cradle of Life on Earth

The birth of human civilization is not an accidental coincidence, but deeply rooted in the earth's unique and intricate natural environment, and in the biological evolution of billions of years. The earth is like an elaborate work of art, with the right temperature, akin to a mother's warm embrace, fostering the hope of life; the flow of liquid water providing the vitality for life, and the perfect atmosphere acting as a strong shield, protecting all creatures from the harshness of space.

Back to the beginning of the birth of the earth, it was a hot and chaotic scene, with magma swirling like a turbulent torrent, and meteorites frequently hitting like dense rain. However, time is like a magical painter, slowly sketching the contours of life. As the eons passed, the earth gradually cooled, and the original oceans began to take shape. This mysterious "soup of life" was like a vast chemical laboratory, where various chemical components collided and blended in a jubilant dance. After countless fortuitous encounters, the first single-celled creature emerged, like the dawn, breaking the darkness of life and heralding the commencement of the journey of life. Since then, life has been like a runaway horse, setting forth on an unending evolutionary journey.

In this long and miraculous time tunnel, life forms have been ceaselessly innovating, evolving from the original simplicity to the complex structure and functional multicellular organisms. The journey from the creatures in the sea to those on land is marked by every seemingly small but significant evolutionary breakthrough. These breakthroughs laid the foundation for the building of human civilization and the subsequent vigorous development of the great powers.

For instance, the evolution of fish with lungs allowed them to survive on land, a crucial step towards the emergence of terrestrial life. The development of amphibians from fish was another monumental leap, enabling life to adapt to both aquatic and terrestrial environments. The rise of the dinosaurs, though eventually culminating in their extinction, paved the way for the mammals to dominate the land. The evolution of primates, with their increasingly complex social structures and tool use, set the stage for the development of human intelligence.

The advent of Homo sapiens, with their advanced cognitive abilities, marked a turning point in the narrative of life on earth. They began to manipulate their environment, creating tools, building shelters, and developing language. This led to the creation of agriculture, which in turn allowed for the establishment of permanent settlements and the rise of civilizations. The domestication of animals and the cultivation of crops were revolutionary changes that transformed human societies, enabling them to sustain larger populations and develop complex social hierarchies.

The development of writing systems was another pivotal moment, allowing humans to record history, laws, and knowledge, which could be passed down through generations. This led to the creation of cities, the rise of empires, and the flourishing of arts and sciences. The wheel, the arch, and the written word were among the many innovations that propelled human civilization forward.

As we look back on this incredible journey, it becomes clear that every step in the evolution of life was a critical building block for the development of human civilization. From the formation of the earth's crust to the rise of Homo sapiens, each event was a brushstroke in the grand painting of life's history. The intricate dance of natural selection, genetic variation, and environmental pressures has shaped the world we live in today, a world where human civilization stands as a testament to the resilience and ingenuity of life.

1.1.2 Evolutionary Turning Point: The Ancient Ape Opens the Dawn of Wisdom

In the magnificent tapestry of life's evolution, the progenitor of humanity, the **opopithecus**, discovered a sanctuary amidst the exuberant and dynamic expanse of the African continent's tropical rain forests. This verdant realm, teeming with life, heralded an era of enduring harmony between mankind and the natural world. The ancient apes, robust and full of vigor, thrived in the dense foliage of the forest floor. With their nimble limbs, they foraged for a diet of fresh, succulent fruits and verdant leaves, leaping lightly from branch to branch, climbing with agility, and cleverly evading the pursuit of natural adversaries. They reveled in the bountiful offerings of nature, living a life of relative ease and contentment.

However, as the relentless wheel of time continued its inexorable march, the capricious artist of the environment subtly shifted the strokes of his brush. The Earth entered an epoch of capricious climatic shifts, akin to a sudden, violent storm that shattered the tranquility of the ancient ape's existence. The once vast and unbroken expanse of the forest began to recede like the ebbing tide, and the verdant home of the ancient apes fragmented into isolated patches. Confronted with unprecedented challenges to their survival, the ancient apes were thrust into a critical juncture of life and death. In this moment of existential crisis, the ancient apes demonstrated their resilient capacity for adaptation, and were compelled to embark on a groundbreaking journey of descent from the trees to the ground.

This monumental transformation, akin to a lightning bolt cleaving through the darkened sky, heralded the dawn of human evolution towards sapience. The upright-walking apes, with hands liberated from the constraints of locomotion, seemed to be bestowed with a mysterious magic that endowed them with the power of creation. Initially, they instinctively utilized the stones and branches strewn across the ground for self-defense against foreign predators or to procure sustenance for survival. But as the aeons unfurled, the sagacity of the ancient apes blossomed, and they began to learn the rudiments of tool-making. With their coarse hands, they patiently abraded stones against each other, fashioning sharp edges to create stone axes, stone knives, and other implements. This nascent ability to manipulate their environment marked the genesis of human ingenuity and the gradual ascent of our ancestors from the ranks of mere animals to the threshold of civilization.

1.2 Economic Activity: Material Drive of the Origin of Civilization

1.2.1 Agricultural Revolution: A Leap in Settlement and Food Production

Looking back about 10,000 years ago, an agricultural revolution that profoundly changed the trajectory of human destiny like spring rain, just like the dawn, quietly lit up in many corners of the world. At that time, in the long-term gatherer and hunting life, humans gradually observed the growth rules of plants, began to try to tame wild plants and animals, completely bid farewell to the nomadic life of living homeless and relying on gatherer and hunting, and steadily moved towards a new era of settled farming life.

In the magical land of the Middle East, known as the "crescent moon land", the industrious and intelligent people took the lead in sowing the seeds of wheat, barley and other grains, watching them thrive under the sun, rain and dew. People took good care of the crops, drove away the birds that stole grain, and built simple irrigation ditches to ensure adequate water. At the same time, the tame sheep, goats and other livestock, opened a new chapter of animal husbandry. People have learned how to raise livestock and let them breed in a safe environment, with access to stable meat, milk and fur resources.

And in the far East, China's Yellow River basin and the Yangtze River basin, also staged a magnificent legend of agricultural development. The ancestors of the Yellow River basin, faced with the surging Yellow River water, were not afraid of difficulties, using the fertile loess to develop chestnut for agriculture. They make the exquisite

Lei Qiang and other farm tools, plow the land, sow millet, year after year, harvest the food of the survival. The Yangtze River basin is the cradle of rice cultivation with its warm and humid climate. People transplanting rice and weeding in paddy fields, carefully care for the growth of rice. The cultivation of rice and millet is like a prairie fire, booming, providing an inexhaustible and inexhaustible material source for the breeding and growth of Chinese civilization.

The emergence of agricultural production, like a solid lighthouse, not only makes food supplies stable and reliable, enabling rapid population growth, but also gave birth to a new model of human settlement. People built houses with longing, at first just using branches and soil to build simple houses, and then gradually developed to build stronger houses with stone and wood. They carefully stored the grain, dug the cellars, made clay POTS, and kept the harvested grain properly. Villages are scattered on the earth like stars, and then gather and integrate, and develop into a bustling town. People in the village cooperate with each other, some are responsible for farming, some are good at hunting, and some are specialized in handicrafts. This settlement mode is like an invisible bridge, which greatly promotes the communication and cooperation between people and creates unique conditions for the further prosperity of civilization.

1.2.2 The Germination and Development of Handicraft Industry

As the bedrock of agricultural production gradually solidified, the handicraft industry began to flourish, bursting forth like bamboo shoots after a spring rain. With the bountiful harvests and the accumulation of surplus agricultural products, communities found themselves with an excess of materials. This surplus allowed them to embark on a fascinating journey of manual creation, harnessing the wealth of their natural resources.

They began to mold and shape the earth beneath their feet, crafting pottery from the very soil. The process was rudimentary at first, involving the simple kneading of clay, but it evolved into a sophisticated art form. The pottery was then fired in kilns, transforming it into a durable and practical commodity. These earthen vessels not only met the daily needs of storing grains and cooking meals but also became a canvas for early artistic expression. The pottery was adorned with exquisite painted designs, featuring imaginative geometric patterns and vivid depictions of animals and plants. Some pieces bore the image of running deer and flying birds, capturing the aesthetic preferences and the vibrant spiritual life of the people of that era.

Simultaneously, the textile industry experienced a leap in technological advancement, marking a fareware to the primitive era when humans draped themselves in leaves. The journey from gathering wild plant fibers, such as those from hemp plants, to domesticating silkworms and reeling silk was a significant one. The textile technology became increasingly refined, and the production of clothing transcended its basic function of providing warmth. It evolved into a means of distinguishing social classes and highlighting individual status. The nobility adorned themselves in luxurious silk garments, intricately embroidered with elaborate patterns, while the common folk typically wore simpler linen attire. The burgeoning handicraft industry, acting as a powerful engine of progress, greatly enriched the material lives of people. It effectively promoted the nascent signs of commodity exchange and began to construct a multi-dimensional and three-dimensional society.

1.2.3 The Construction of Trade Network: Resource Circulation and Civilized Exchange

The stability of settlement life and the increasing specialization of production, driven by two engines, make trade more frequent, like a surging tide. The intricate dance of supply and demand has led to a flourishing exchange of goods and services, as people in different regions, akin to intelligent businessmen, capitalize on their unique resource advantages. They create characteristic products with ingenuity, harnessing the natural resources and cultural heritage of their lands. With fearless courage and unwavering determination, they venture into unknown territories, sharing their resource dividends and partaking in the wealth of others.

In the ancient Mediterranean region, the Greek city-states, with their superior navigation conditions, became the masters of the sea. They shuttled freely between the waves, conducting frequent trade with Egypt, Phoenicians, and other distant places. Grain, wood, precious metals, and handicrafts were among the precious materials that flowed through their bustling ports. The olive oil and wine from Greece were particularly popular due to their excellent quality, while Egyptian papyrus and grain met some of the needs of the Greek city-states. Merchant ships from both sides traded in the Mediterranean, the crew battling the wind and waves, as they transported local specialties to distant lands.

The famous Silk Road, just like a gorgeous ribbon, firmly connects Asia, Europe, and the African continent. Chinese silk, with its smooth and delicate texture and bright colors, has become a coveted luxury for western aristocrats. Starting from Changan, the caravan, laden with silk, porcelain, tea, and other exquisite specialties, was exported to the west along this legendary road. They traversed the vast desert, surmounted steep mountains, and endured many hardships, all in exchange for exotic customs, rich spices, dazzling jewelry, vigorous and brave good horses, and other rare treasures. Oasis cities along the way, such as Dunhuang and Loulan, prospered because of trade and became important nodes of cultural exchange, where ideas, religions, and philosophies mingled with goods, enriching the civilizations that lay beyond the horizon.

The careful construction of the trade network has not only realized the efficient distribution of goods but has also woven a tapestry of human interaction, fostering understanding and cooperation among disparate cultures. This intricate web of commerce has been the lifeblood of civilizations, enabling them to thrive and grow, and has laid the groundwork for the interconnected world we know today.

1.3 Political Organization: The Construction Framework of Civilized Order

1.3.1 Tribal Society: Primal Collaboration and Order

In the nascent stages of human civilization, the dawn of our existence saw us living in close-knit communities, where tribes formed the bedrock of our social structure. These tribes, typically composed of anywhere from a few dozen to several hundred individuals, established their homes in areas characterized by dense forests, fast-flowing rivers, or expansive grasslands. These natural surroundings not only provided a habitat but also served as a reservoir of resources essential for the tribe's sustenance and survival.

Within these tribal communities, the social fabric was woven with threads of simplicity and directness. Leadership often fell to one or more individuals who had earned the respect and trust of their peers through their wisdom, experience, and strength. These leaders, akin to sagacious navigators steering their people through the uncharted waters of existence, were frequently the most formidable, courageous, or astute members of the tribe. They bore the weighty responsibilities of organizing hunting expeditions, defending against external threats, and ensuring equitable distribution of food. Their exceptional prestige and adept leadership skills were instrumental in upholding the tribe's harmonious order and collective well-being.

Hunting, a critical activity for survival, was conducted with strategic precision under the guidance of these leaders. They leveraged their deep understanding of animal behaviors to devise effective hunting strategies, whether it was setting elaborate traps or orchestrating group drives to corner and capture prey. Once the hunt was successful, the leader would distribute the spoils with fairness, taking into account the needs and contributions of each tribal member. This ensured that everyone received a portion sufficient for their survival, fostering a sense of unity and shared purpose among the tribe.

The members of the tribe worked in unison, much like soldiers in a battle, hand in hand, to surmount the myriad challenges that lay in the path of their existence. This cooperation cultivated a spirit of pure and unadulterated primitive collectivism, where the welfare of the group took precedence over individual desires. This social bond, deeply rooted in both blood relations and the shared need for survival, formed a solid foundation that laid the groundwork for the emotional and practical underpinnings of future political organizations.

As children grew within the tribe, they were nurtured in an environment where they learned essential survival skills from their elders. They absorbed the wisdom of their ancestors, inheriting the culture, traditions, and knowledge that had been passed down through generations. This cultural transmission was not merely an academic exercise but a living, breathing process that ensured the continuity of the tribe's identity and way of life. The cohesion and solidarity of the tribe were further strengthened by these shared values and practices, creating a resilient community capable of withstanding the tests of time and nature.

1.3.2 The Rise of Chiefs: Centralization of Power and Social Stratification

As the population grew and the competition for resources became increasingly fierce, some tribes have undergone a magnificent transformation and have gradually grown into chiefs. This transformation was not merely a change in leadership but a profound shift in societal structure. Compared with the relatively egalitarian nature of tribal societies, the power structure of the chiefs is like a towering pyramid, more centralized and stable, with the leaders standing at the peak of the power, wielding greater decision-making power and the crucial right to allocate resources. This centralization of power allowed for more efficient management of the growing populations and the complex issues that arose from increased competition for resources.

The emergence of chiefs was often accompanied by further development of agricultural production. As agricultural techniques improved, regions were able to produce more food and feed larger populations. This surplus allowed for the sustenance of more people beyond those directly involved in food production, leading to the development of specialized roles and professions. Social structures began to become complicated when a region was able to produce more food and feed more population. The surplus also meant that some individuals could accumulate wealth and power, leading to the stratification of society.

At the same time, the social structure, like a multi-prism, began to show a stratification phenomenon. In addition to the vast number of ordinary people, who were primarily involved in agriculture and other basic labor, society began to differentiate into various classes such as the noble aristocracy, the mysterious priest class, skilled artisans, and merchants. The nobles are like the right arm of the leader. They are usually the relatives of the leader or those with outstanding performance in war and production, assisting the leader to manage daily affairs in an orderly way. They were responsible for organizing labor force for large-scale farmland reclamation, water conservancy construction, and other communal projects that were vital for the sustenance and growth of the community.

Priests are like the messengers of the gods, responsible for devout and solemn religious sacrifice activities, praying for the prosperity of the tribe, peace, and the well-being of its members. They have a certain religious knowledge and mysterious rituals, and have a high prestige in the hearts of the tribal members. Their role was not only spiritual but also social and political, as they often served as advisors to the chiefs and played a crucial part in maintaining social cohesion and order.

The emergence of the chieftain marks a key step in the initial concentration of political power. The social order has become more and more complex and diversified under the fine norms of the hierarchy, which has accumulated valuable experience for the formation of a more advanced political form of the state. This concentration of power and the development of complex social structures laid the groundwork for the eventual emergence of civilizations with more sophisticated governance systems, legal frameworks, and social hierarchies.

1.3.3 The Birth of a Country: Coercion and Public Administration

When the productive forces are further developing rapidly like a rocket, the social division of labor is like a precise gear group, increasingly delicate and complex, and the country arises at the historic moment in the call of history. With the powerful public power as the solid backing, like a majestic giant, through the establishment of a daunting army, the formulation of rigorous and fair laws, the collection of reasonable taxes and a series of powerful means, carefully maintain social order, and protect the public interests.

As the engine of society accelerates, the intricate machinery of division of labor becomes more sophisticated, mirroring the complexity of a finely tuned clock. Each gear, each cog, represents a specialized role within the vast network of production and services. The state emerges as a necessary entity, a guardian of order and progress, in response to the growing needs of society. With its robust public power as a foundation, the state stands tall, akin to a colossus, establishing a formidable military presence to deter threats and enforce peace. It crafts laws that are both stringent and equitable, ensuring justice and fairness prevail. The state collects taxes not arbitrarily, but with a sense of responsibility, funding essential services and infrastructure that benefit the populace. Through these and other potent mechanisms, the state vigilantly upholds the social fabric, safeguarding the interests of the public.

The mysterious and brilliant pharaoh dynasty in ancient Egypt, standing on the Nile river, the pharaoh was regarded as the embodiment of the sun god, with the supreme power. He controls a vast bureaucracy, ranging from agricultural irrigation officials to ministers overseeing construction projects. In order to ensure agricultural production, the Egyptians built magnificent pyramids, which were not only the tomb of the Pharaoh, but also the large-scale labor force of the state and the use of advanced engineering technology.

The ancient Egyptian civilization, shrouded in the mists of time, was a beacon of human achievement. The pharaoh, seen as a living god, presided over a society that was both hierarchical and highly organized. The bureaucracy he commanded was vast and intricate, with officials ranging from those managing the life-giving waters of the Nile to those overseeing grand construction projects like the pyramids. These monumental structures were not merely the final resting places for the pharaohs; they were also a testament to the state's ability to mobilize and manage a large workforce. The construction of the pyramids showcased the Egyptians' mastery of engineering and their capacity to undertake massive projects, which in turn ensured the stability and prosperity of agricultural production, the backbone of their economy.

Ancient Babylon has the legendary code of Hammurabi kingdom, the code of Hammurabi is engraved on the stone pillars, showing the world the majesty of the law. The code covers all aspects of social life, from commercial transactions to marriage and family, from personal injury to property disputes. It clearly stipulates the code of conduct, severely punishes criminal acts, and ensures the stable operation of the great ship of society to brave the wind and waves.

The Code of Hammurabi, one of the earliest and most complete legal codes in human history, stands as a testament to the sophistication of ancient Babylonian society. Inscribed on towering stone pillars, the code was a public declaration of the rule of law, a concept that would echo through the ages. The code's provisions touched every facet of life, providing a framework for commerce, defining the rights and responsibilities within family structures, and setting standards for compensation in cases of personal injury or property damage. It laid out a clear set of rules and consequences, ensuring that justice was served and order maintained. This legal foundation allowed Babylonian society to navigate the turbulent waters of human interaction with a steady hand, fostering an environment where trade could flourish and civilization could advance.

The Xia, Shang and Zhou dynasties, which have a long history in ancient China, are all outstanding classical representatives of the early state form. As the first dynasty in Chinese history, the Xia Dynasty opened the hereditary system and broke the original abdicati.

The Xia, Shang, and Zhou dynasties of ancient China represent the formative years of statecraft and governance in the region. The Xia Dynasty, considered the first dynasty in recorded Chinese history, established the hereditary system, a practice that would become a cornerstone of Chinese imperial rule for millennia. This system allowed for a stable transfer of power, reducing the chaos that might arise from contested successions. The Shang Dynasty, which followed, further refined the concept of a centralized state, with a king at its apex, supported by a complex administrative system. The Zhou Dynasty, the longest-lasting of the three, saw the development of the Mandate of Heaven theory, which posited that a ruler's right to govern was granted by divine will. This concept provided a philosophical basis for the legitimacy of rulers and the justification for rebellion when a ruler lost the mandate. These early dynasties laid the groundwork for the bureaucratic and philosophical traditions that would characterize Chinese civilization for centuries to come.

1.4 Cultural Achievement: The Casting Light of the Soul of a Civilization

1.4.1 Language: A Tool for Information Inheritance and Thought Exchange

Language, akin to a radiant dawn, stands as one of the pivotal markers that set human beings apart from the rest of the animal kingdom. In the grand odyssey of human civilization's genesis, language plays an indispensable and foundational role. Early humans, much like innocent children, communicated their hunting expertise through rudimentary and straightforward language, passing on survival techniques as if passing on the flame of knowledge. This exchange fostered group collaboration and enabled humanity to advance steadily along the path of existence.

Within the confines of primitive tribes, individuals would gather around the communal bonfires, engaging in conversations through simple sounds and gestures. They shared the fruits of their daily hunts and deliberated on strategies to confront the ferocious beasts that threatened their survival. As these tribes evolved and grew, their language expanded in complexity, enabling them to articulate more intricate emotions and thoughts.

As the wheel of society turned and progressed, the creation of words emerged as a revolutionary force, comparable to a seismic shift that altered the course of history. This was a monumental breakthrough with epoch-making significance. The cuneiform script, a product of Sumerian ingenuity, was initially conceived to document commercial transactions, land distribution, and other administrative matters. It involved the use of reed stalks to etch wedge-shaped symbols onto clay tablets. Once dried or fired, these tablets could endure through the ages, offering invaluable historical records for future generations to study and learn from.

The enigmatic hieroglyphics of ancient Egypt, adorned with vivid and intricate patterns, depicted deities, animals, and human figures. These symbols were not only integral to religious ceremonies and sacrifices but also served to chronicle the accomplishments of pharaohs and the lineage of dynasties. The profound and ancient Chinese oracle bone inscriptions, etched onto tortoise shells and animal bones, primarily served a divinatory purpose. By interpreting the cracks that appeared on these oracle bones, soothsayers could predict events of good fortune, misfortune, and other aspects of destiny. This practice was deeply embedded in the cultural and spiritual fabric of ancient Chinese society, influencing decision-making and governance.

These early forms of writing not only facilitated communication and record-keeping but also laid the groundwork for the development of complex systems of governance, law, and culture. They enabled the preservation of knowledge across generations, allowing societies to build upon the wisdom of their ancestors. The evolution of language and writing systems stands as a testament to human ingenuity and our insatiable quest for understanding and connection.

1.4.2 *Religious Belief: The Link Between Spiritual Sustenance and Social Cohesion*

In the face of the mysterious and powerful forces of nature, and the ubiquitous uncertainty in the process of survival, religious belief is like a warm lamp, quietly lit in the human heart. Early humans, much like devout believers of today, worshiped the natural gods, such as the shining sun god, the cold and clean moon god, the violent and unruly wind god, and the nurturing and life-giving water god. Through the solemnity of sacrificial ceremonies, they prayed to these deities for good weather, bountiful harvests, and protection for their tribes. These rituals were not just a means of supplication but also a way to make sense of the world around them, to find meaning in the cycles of day and night, the changing seasons, and the unpredictable forces of nature.

In ancient Egypt, the sun god Ra was regarded as the creator and master of all things. The Pharaohs, as the earthly representatives of Ra, declared themselves to be descendants of the god and possessed divine lineage, which served to strengthen their ruling authority. The grandeur of the sun temple in Heliopolis was a testament to the importance of the sun god in Egyptian culture. Every day, priests and people alike participated in grand sacrificial ceremonies, offering a variety of sacrifices, including cattle, sheep, grain, and wine, all in the hope that the sun god would bless the Nile with abundant water and ensure a good agricultural harvest.

Religious belief not only provides people with spiritual comfort and a sense of security in their hearts but also acts like a potent medicine to relieve fear, soothing the deep-seated fear of the unknown world. It also acts like a sturdy rope, becoming an important bond that unites social forces. Inspired by the sacred and solemn spirit of religious ceremonies, tribal members would come together as one, imitating the integrated strength of a steel flood, participating in sacrifices and the building of temples. This collective action constantly reinforced their collective identity and sense of community.

The distinctive teachings and rituals of different religions, like colorful brushes, permeate into all aspects of social life, carefully shaping people's values and morals. These teachings have a profound and lasting impact on the development of civilization, just as bright stars illuminate the path of human civilization. The idea of compassion and universality, for instance, influenced the development of societies, promoting concepts of community and shared humanity. In ancient India, Buddhism spread the message of non-violence and the interconnectedness of all life, which shaped the social fabric and influenced governance and social structures. In medieval Europe, Christianity's teachings on charity and love for one's neighbor helped to establish hospitals and care for the poor, reflecting the religion's impact on societal values and practices.

Religious belief has been a driving force in human history, influencing art, architecture, literature, and even science. The grandeur of Gothic cathedrals and the intricate designs of Islamic art are testaments to the influence of religious aesthetics. The pursuit of knowledge and understanding in the Islamic Golden Age was fueled by a religious imperative to understand God's creation. In modern times, religious movements continue to shape the world, from the civil rights movement in the United States, inspired by the teachings of Martin Luther King Jr., to the peace activism of Mother Teresa in India.

Thus, religious belief has been a constant companion to humanity, offering comfort, guidance, and a framework for understanding the world. It has been a source of inspiration for countless individuals and communities, driving them to achieve great things and to strive for a better, more compassionate world.

1.4.3 Artistic Creation: The Sublimation of Aesthetic Pursuit and Emotional Expression

Artistic creation is like a blooming flower, revealing fragrance in the garden of human civilization. It is not only the persistent pursuit of beauty, but also the surging release of emotion. From the rough but shocking rock paintings in the primitive cave, art is closely linked with human civilization, walking hand in hand.

In the Lasco Cave in France, the inner wall is painted with lifelike animal images such as bison and wild horses. These rock paintings are carefully drawn by early humans with mineral pigments. The lines are simple and powerful, but they accurately capture the clever posture of animals running and jumping. Creators with careful observation of animals and the fear of deep heart, their eternal freeze on the wall, perhaps to record a successful hunting, or pray for good luck, these paintings like a window to the ancient times, let posterity to explore the human life and spiritual world.

With the continuous evolution of civilization, architectural art has gradually emerged and become a towering monument of human wisdom and strength. The ancient Greek Parthenon, with its elegant and solemn style of multiple pillars and exquisite golden ratio, shows the ultimate pursuit of the beauty of harmony and order. The sun shines on the white marble column, and the light and shadow intersect, creating a sacred and solemn atmosphere. It is not only a temple dedicated to the gods, but also the lofty praise of the ancient Greeks to the ideal world and the perfect human nature, carrying the glory and faith of the city.

In the east, the ancient Chinese palace buildings are also magnificent. The Imperial Palace in Beijing has red walls and yellow tiles, and the eaves are like the wings of birds, showing the majesty and dignity of the royal family. The layout of the palace strictly follows the principle of symmetry of the central axis. Every brick, every beam and every pillar embody the painstaking efforts of the craftsmen, contain profound cultural deposits, reflect the philosophy of "the unity of the man of nature", tell the rise and fall of the dynasty, and witness the torrent of history.

In the field of painting, there are many talents and excellent works. Renaissance masters, such as Da Vinci, Michelangelo, Raphael, broke the darkness and imprisonment of the Middle Ages with their brushes. Da Vincis Mona Lisa, the mysterious smile like thousands of words, behind the delicate insight into human nature; Michelangelo in the ceiling of the church, character muscles tight, passionate, show unparalleled vitality and creativity, as if the first moment of human birth; Raphael is gentle and kind, full of maternal glory, with love and beauty of the world mind. Their works not only pursue the visual pleasing and pleasing to the eye, but also contain the profound exploration of the humanistic spirit, add bright treasures to the treasure house of human art, and promote the civilization to a higher level of aesthetic realm.

Music, as an art form spanning the language, is like a clever stream, nourishing the human soul. From the simple drums and ballads of the primitive tribes, used to celebrate the harvest and drive away the wild animals, to the magnificent movements created by Mozart, Beethoven and other masters in the classical music period, the connotation and expression of music are continuously expanded and sublimated. Mozarts symphony is light and lively, like the warm spring sun, full of optimism and innocence; Beethovens Symphony of Fate, the passionate melody is like the roar of the tenacious struggle against fate, giving countless people the courage and strength to forge ahead in difficulties. Music can arouse the most sincere emotional resonance in peoples hearts, cross the boundary of time and space, let people from different regions and times find the comfort and sense of belonging in the melody, and become an indispensable spiritual nourishment for human civilization.

Literary works are like a mirror, reflecting the joys and sorrows of human society. From the ancient epic "Iliad" and "Odyssey", the heroes of honor, courage and wisdom, to the Chinese Book of Songs, with the simple poem of labor, love, war and other life scenes, "Jujiu cage, in the river, the fair lady, gentleman qiu", flowing with the pure and beautiful; to the modern novels such as Tolstoys "war and peace", shows the human complex, social changes, thought-provoking. Literature takes words as bricks and stones, constructing a rich and colorful spiritual world, in which readers can roam and imagine, inherit the essence of human thought, and weave a brilliant chapter for the continuation and development of civilization.

In short, artistic creation throughout the whole process of human civilization development, it with unique charm inspired human yearning for beauty, become the emotional haven and thought incubator, let human in the pursuit of material progress at the same time, do not forget to protect mind poetic habitat, inject vigor and vitality, for civilization make it bloom the light of never fade.

To sum up, the origin of human civilization is a magnificent epic of the interweaving and synergy of natural environment, biological evolution, economic activities, political organization and cultural achievements. The gift of the earth provides a hotbed for the origin of life, the spark of biological evolution to human wisdom, economic activities build a solid material foundation, political organization builds the framework of order, and cultural achievements light up the light of civilization. In the long history of China, these elements influence and complement each other, promoting human beings from ignorance to civilization, from barbarism to civilization, and continue to write brilliant chapters. In the future, driven by the two wheels of inheritance and innovation, human civilization will forge ahead towards a more brilliant distance, constantly explore the boundary of the universe, humanity and the unknown, and continue to write the legend of all mankind.

Chapter 2
The Development of Human Civilization

In the vast and mysterious universe, the Earth stands out as a blue gem, shining with its unique light, carrying the splendid and unique great creation of human civilization. From the ancient humans of the primitive society, who lived in caves and hunted for survival, enduring countless hardships, facing the elements, and braving the wind and rain, the human journey has been one of relentless perseverance and adaptation. Over millennia, these early societies developed agriculture, which allowed for the establishment of permanent settlements and the rise of civilizations.

Through the ages, empires rose and fell, each leaving behind monumental architecture, art, and literature that still inspire awe today. The Great Pyramids of Giza, the Great Wall of China, the Hanging Gardens of Babylon, and the Colosseum of Rome are just a few examples of the architectural prowess that ancient civilizations achieved. These structures not only demonstrate the ingenuity of our ancestors but also their capacity for organization and cooperation.

As time marched on, the Renaissance period in Europe brought about a rebirth of art, science, and philosophy, challenging the status quo and ushering in a new era of human thought. Figures like Leonardo da Vinci and Michelangelo left an indelible mark on the world with their masterpieces, while thinkers like Galileo and Copernicus revolutionized our understanding of the cosmos.

From ancient times to the present, the development of human civilization has been a journey of continuous evolution and adaptation. The Industrial Revolution marked a significant turning point, transforming societies from agrarian to industrialized nations. This period saw the rise of machines, factories, and the modern cityscape, which in turn led to the development of new social classes and the redefinition of work and life.

In the twentieth century, the world witnessed two world wars that reshaped the geopolitical landscape and the emergence of the digital age, which has connected the globe like never before. The internet, smartphones, and social media have become integral parts of daily life, enabling instant communication and access to information on a scale unimaginable to previous generations.

L. Yan, *Value-Based Civilization*, Understanding China,
https://doi.org/10.1007/978-981-96-9490-7_2

The evolution of human civilization, like a spectacular, exciting grand journey, has left countless deep imprints full of human wisdom and courage. In this long and tortuous course of development, values have always been like a tenacious rope, closely connecting all stages. They have been the guiding principles that have helped societies navigate through times of change and challenge. From the pursuit of justice and equality to the importance of freedom and democracy, these values have been the bedrock upon which human societies have built their institutions and laws.

As we stand on the cusp of the future, the key force that continues to promote human society to forge ahead and climb new peaks is the relentless pursuit of knowledge, the spirit of innovation, and the unyielding belief in the potential of humanity. The story of human civilization is one of resilience, creativity, and an unquenchable thirst for progress, a narrative that continues to unfold with each passing day.

2.1 Primitive Society: The Germination of Civilization, the Initial Value

Back millions of years ago, human civilization was still in its nascent stages, emerging amidst the tumultuous backdrop of a primitive society. Our ancestors, equipped with nothing more than the most rudimentary tools, grappled with an environment that was both unforgiving and unpredictable. The world they inhabited was a wild expanse, teeming with untamed beasts and subject to the whims of a capricious climate. Survival was not just a priority; it was the sole, pressing concern that dictated every moment of their existence. In the long and arduous years that stretched before them, people banded together, relying on the most basic survival instincts to eke out an existence. Day after day, they ventured forth, hunting for sustenance, gathering wild fruits, and searching for precious water sources, all while teetering on the precarious edge of life and death. Their sole aim was to ensure the continuation of their lineage.

Even amidst such adversity, the seeds of value and culture were quietly taking root. Matters that were intimately connected to the survival of the tribal members were accorded the highest importance. The food procured through hunting was not merely sustenance; it was the very essence of life, with each morsel representing a renewed chance at enduring the harshness of their world. The natural caves that offered shelter from the elements, storms, and protection from wild beasts, became sanctuaries that secured the tribe's very existence. As time marched on, the spark of human ingenuity began to ignite, and gradually, our ancestors mastered the art of tool-making. They crafted carefully polished stone axes and fashioned sharp, pointed bone spears. These seemingly crude implements were the embodiment of human creativity and ingenuity. They dramatically enhanced the efficiency of hunting and gathering, marking the initial strides towards civilization. Their significance extended far beyond the tools themselves; they were a testament to humanity's determination to challenge the forces of nature and assert control over their own destiny.

Simultaneously, within the primitive society, people began to develop simple yet pragmatic social norms and values. Tribal members learned to work in unison and to share food 无私ly, for they understood that only through the unity of the group could they hope to withstand the brutal culling imposed by nature. This mutual assistance, born out of the necessity to survive, was akin to the first stirrings of a social contract. It was a tacit understanding that the collective strength of the tribe was greater than the sum of its individual parts. The bonds formed through these shared struggles and triumphs laid the foundation for the complex social structures that would eventually define human civilization. As the tribe grew and prospered, so too did the richness of their culture, with stories, traditions, and rituals emerging to bind them together and give meaning to their existence. Each generation built upon the knowledge of the last, passing down survival skills and wisdom, ensuring that the flame of human progress would never be extinguished.

2.2 Agricultural Revolution: The Foundation of Civilization, Value Expansion

About 10,000 years ago, an epoch-making agricultural revolution, like a surging tide, completely changed the course of human life and laid a solid foundation for the prosperity of civilization. Human beings began to try to domesticate animals and plants, carefully planted food crops, raised livestock, gradually bid farewell to the hunter-gatherer life without settled abode and completely dependent on natural gifts, and steadily moved towards the agricultural production mode of settlement.

In this brand-new agricultural society, fertile land has suddenly become the most precious wealth in peoples eyes, and is the core symbol of the foundation of survival. Year after year, spring planting and autumn harvest, people sweat on the land, intensive cultivation, accumulated a rich and practical agricultural knowledge and exquisite technology. At the same time, the handicraft class rose rapidly, making exquisite pottery, fabrics and other daily necessities by virtue of the exquisite skills passed down from generation to generation or studied by themselves. These handicrafts, on the one hand, meet the actual needs of daily life, is the people; on the other hand, they carry the unique cultural connotation, with the unique artistic beauty, express the aesthetic pursuit, the value from the material practical level to the spiritual and cultural level.

And the invention of words is undoubtedly an earth-shattering value change in the agricultural era. It is like a magic master key, breaking the narrow limitation that knowledge can only be passed on from mouth to mouth, so that the experience and profound and complex thoughts accumulated by generations of mankind can be crossed across the long time and space barrier and passed on to future generations. Since then, philosophy, literature, science and many other fields have been like flowers in spring, opening a glorious course of vigorous development, and human spiritual civilization has blossomed with unprecedented light. In the birthplace of ancient and mysterious civilization, like the Sumerian civilization in the Nile River basin and the ancient Egyptian civilization, writing is used in widely and deeply records astronomy and calendar, to help people grasp the farming time and predict the celestial phenomena; to write religious teachings, to provide spiritual sustenance and belief guidance; to compile laws and regulations, to maintain the stable operation of social order. The emergence of words provides an indispensable strong support for the establishment of human social order and the long history of culture, which is an eye-catching symbol of the maturity of human civilization.

With the vigorous development of agriculture, the structure of social organization is also increasingly complex and delicate. Cities, a brand-new gathering place of civilizations, are gradually emerging. The city attracts a large number of people from all directions, and people with different identities and different skills gather here, giving rise to a more detailed and diversified social division of labor. The ruling class is responsible for controlling the overall situation and making the policies and regulations; the priests bear the responsibility of communicating the gods and protecting the spirit; the craftsmen make practical and beautiful objects with the dexterous hands to promote the flow of goods. Each group performs their own duties and cooperate with each other, just like all the parts in a precision instrument, jointly pushing the great ship of civilization rolling forward. It is accompanied by the continuous deepening and expansion of a series of values. Peoples pursuit of social order is increasingly persistent, the awe of the gods is evident, and the importance of family inheritance is also integrated into the daily life. These concepts, like the invisible but tough silk thread, keep the agricultural society closely together and ensure the stable operation and sustainable development of the society.

2.3 Classical Civilization: Diverse Bloom, the Value Shines

With the continuous development of human society, classical civilization has risen in different regions one after another, like a colorful wonderful flowers, presenting a diverse and brilliant scene.

In the ancient east of China, the Chinese civilization stands out with its profound heritage and unique and charming charm. The five core concepts of "benevolence, righteousness, propriety, wisdom and faith" advocated by Confucianism, just like five dazzling stars, run through every tiny corner of social life, and become the fundamental value principle for Chinese children to pass them down from generation to generation.

As the core essence of the Confucian ideological system, "benevolence" is derived from the purest kindness of the deepest heart, advocating promoting oneself and others, caring for others and pursuing the harmonious coexistence of society. Within the family, respecting the elderly and the daily embodiment of "benevolence", which maintains the close unity among family members and guarantees the inheritance of the family lineage, the rulers live and strive to realize "benevolence", so as to build a stable and prosperous social order."Justice" emphasizes to act impartially and in line with moral principles, especially in the complex field of business contacts, it requires people to stick to the bottom line of integrity, never make ill-gotten gains, build a solid foundation for economic ethics, and ensure the steady operation of business activities on the track of fairness and justice. "Gift" meticulously regulate the interpersonal behavior, from solemn, carrying the national belief of sacrifice ceremony, to the streets, the neighborhood daily greetings, from gesture, smile all show the grace of Chinese civilization, after the baptism of 1000 years, inheriting the Chinese nation unique etiquette culture, become the Chinese nation from other national spirit.

In the West, the ancient Greek civilization is like a dazzling star that breaks through the night sky, illuminating the long road of the subsequent development of western civilization with its light. The persistent pursuit of reason is undoubtedly the value benchmark of the ancient Greek civilization. From Taylor bold boldly explore everything, pioneering open rational speculation, to Socrates in the busy square, through a thought-provoking dialogue, inspire people think deeply about the essence of truth, virtue, and to Aristotle with its extraordinary wisdom and profound knowledge, build up covering philosophy, natural science, political science, and many other fields of huge knowledge system, the darkness of rational light like a sword dispersed in the darkness of human head, for the vigorous development of western academic thought laid a very solid tone.

This spirit of respect for reason naturally permeates into every inch of city-state politics. Citizens actively participate in public affairs, and through democratic ways such as open debate and voting, making decisions on major issues related to the future trend of the city. The prototype of the democratic system is also initially revealed in this ideological collision and practical exploration. In the field of artistic creation, the ancient Greek sculpture makes the precise human body proportion and the perfection of human beauty and rational order shape to bring the rational aesthetics perfectly, setting up the lofty benchmark for the future western art and becoming the source of western aesthetic spirit.

At the same period, the ancient Indian civilization, ancient civilization also have their bloom a dazzing brilliance, according to their own unique religious beliefs, the special geographical environment and specific social development needs, breeds different styles, unique value system, covering religious piety, astronomical calendar, architectural art, and many other fields, the bright civilization achievements like the stars in the night sky, together for the vast treasure of human civilization added endless treasures, greatly enriched the diversity of human civilization and massiness.

2.4 Medieval Times: Inheritance and Twists and Turns, the Value Is Looming

In the Middle Ages, Christianity was not just a religion; it was the very fabric of society, permeating every aspect of daily life with its profound influence. From the grand cathedrals that towered over the skyline to the smallest village chapel, the church's presence was ubiquitous and commanding. It was a colossal and awe-inspiring entity, holding sway over vast tracts of land, immense wealth, and the most precious knowledge resources of the time. The church's doctrines and dogmas were the bedrock upon which the moral and ethical foundations of medieval society were built, imposing stringent norms and constraints on the thoughts and actions of the populace.

Under the guidance of religious teachings, the people of the Middle Ages fervently pursued the redemption of their souls. The reverence and devotion to God were as integral to their existence as the air they breathed, infusing every moment of their lives with a sense of the divine. The monastery, a sanctum of learning and knowledge, was where monks painstakingly copied classical texts by the flickering light of ancient lamps. This seemingly monotonous and repetitive task was, in fact, a vital endeavor that preserved countless ancient texts for future generations. The monks' dedication ensured that the wisdom of the ancients was not lost to the ravages of time.

The faint flame of the Ming Dynasty laid a precious foreshadowing for the Renaissance that later shook Europe. This period of enlightenment, sparked by the rediscovery of classical antiquity, heralded a new era of intellectual and artistic growth that would fundamentally alter the course of Western civilization.

However, it has to be admitted that this stage of European social development was relatively slow, hindered by the weight of feudal hierarchy that pressed upon the people like a heavy mountain, making it difficult for them to breathe freely. The interplay between secular power and religious authority created a complex web of control that, to a certain extent, stifled the ideological vitality and innovative potential of the populace. Despite these constraints, the progress of human civilization did not come to a complete halt. In the field of agricultural technology, for instance, there were significant advancements such as the improvement of new farm tools and the optimization of irrigation systems, which enhanced productivity and supported the growing population.

The unique and mysterious religious aesthetics and exquisite decoration of Gothic architecture became the distinctive symbol of medieval European civilization. The soaring spires and intricate stained glass windows of Gothic cathedrals were not only places of worship but also served as a testament to the persistent pursuit of belief and beauty in that era. These architectural marvels were a physical manifestation of the people's yearning for the divine and their quest for artistic expression.

In the distant east, China experienced its own remarkable developments. The Song Dynasty, for example, saw significant advancements in science, technology, and culture. The invention of gunpowder, the compass, and printing technology had profound impacts not only on China but on the rest of the world. The agrarian society of China was also marked by the development of sophisticated irrigation systems and the improvement of agricultural tools, which contributed to the prosperity of the empire. The philosophical and literary achievements of the era, including the works of Confucian scholars and poets, were celebrated throughout the land, reflecting a society that valued knowledge and the arts.

The exchange of ideas and culture between the East and West during this period was minimal, yet both civilizations were undergoing transformations that would shape the future of humanity in profound ways. While Europe was deeply entrenched in its religious doctrines, China was experiencing a renaissance of its own, with the flourishing of Neo-Confucianism and the arts. Despite the vast distance separating them, both societies were moving towards a future where knowledge and innovation would play increasingly central roles.

2.5 The Renaissance and the Enlightenment: Awakening and Remodeling, Value Innovation

From the fourteenth century to the seventeenth century, a warm and shocking Renaissance movement, akin to the spring breeze, quietly swept across the European continent. This cultural rebirth, a breath of fresh air, blew away the cobwebs of ignorance and superstition that had long imprisoned the humanistic spirit, suffocated by the rigid constraints of religious theology. It was as if people were awakening from a long winter's slumber, shaking off the frost of dogma to re-examine the splendid cultural heritage left over from ancient Greece and Rome. With renewed vigor, they began to explore and celebrate the intrinsic value, dignity, and potential of human beings, fostering a new and subversive understanding of human nature.

Artists, inspired by this awakening, turned their eyes away from the heavenly realms and celestial beings that had long been the subjects of their canvases. Instead, they found beauty and inspiration in the ordinary people around them, using their lives and stories as the blueprint for their creations. The result was a proliferation of lifelike paintings, rich with emotional depth and realism. Leonardo da Vinci's Mona Lisa, with her enigmatic and beguiling smile, became emblematic of this era. Her expression was like a dawn breaking through the darkness, revealing the complex and beautiful tapestry of human nature. The brushstrokes of masters like Michelangelo and Raphael captured the very essence of humanity, imbuing their works with a sense of grandeur and emotion that resonated deeply with viewers.

Writers, too, embraced this humanistic ethos, expressing personal sincere feelings and offering sharp criticism of the social realities of their time. Boccaccio's "Decameron," for instance, was like a sword that boldly broke the heavy shackles of religious asceticism. It vividly portrayed the multifaceted and colorful aspects of humanity, with stories that ranged from the humorous to the tragic, from the mundane to the profound. This literary work, along with others like Shakespeare's plays and

Cervantes' "Don Quixote," not only entertained but also challenged the status quo, encouraging readers to think critically about their society and their place within it.

The Renaissance was not merely an aesthetic revolution; it was a comprehensive transformation that extended into the realms of science, philosophy, and politics. Figures like Nicolaus Copernicus and Galileo Galilei dared to challenge the geocentric model of the universe, laying the groundwork for modern astronomy. The humanist scholars of the Renaissance rediscovered and translated ancient texts, fostering a revival of classical learning that would fuel the intellectual growth of Europe for centuries to come.

This movement, like a raging fire of the liberation of the mind, completely burned through the shackles of religious theology that had bound people's thoughts since the Middle Ages. It made the brilliance of humanity shine again in the European continent, strongly promoting the vigorous development of art, literature, science, and many other fields. The Renaissance opened a new chapter in the continental modern civilization, setting the stage for the Enlightenment that followed.

Following the excitement of the Renaissance, the Enlightenment movement reached its climax like a turbulent tide in the eighteenth century. Thinkers, like warriors holding high the torch of reason, firmly grasped the banner of idealism. They mercilessly criticized the decadent darkness of the feudal autocratic system and the absurdity and ignorance of religious superstition. Philosophers such as Voltaire, Rousseau, and Kant loudly advocated for a series of revolutionary ideas that would fundamentally alter the course of human history. They championed the concepts of individual liberty, freedom of expression, and the pursuit of happiness, ideas that would eventually inspire the American and French Revolutions.

The Enlightenment was a period of intense intellectual and philosophical activity, where reason and scientific inquiry were elevated to new heights. It was a time when the boundaries of human knowledge were pushed forward, and the foundations of modern secular society were laid. The Enlightenment thinkers believed in progress, in the ability of humanity to improve its condition through the application of reason and the pursuit of knowledge. Their ideas spread like wildfire, igniting a passion for reform and change that would sweep across Europe and beyond, forever altering the way we understand the world and our place within it.

2.6 Industrial Revolution: Take-off and Reconstruction, Value Change

In the middle of the eighteenth century, the rolling tide of the industrial revolution hit with an overwhelming trend, completely subverting the traditional mode of production. The era of steam and mechanization swept across continents, transforming societies from agrarian to industrial. Machine production replaced manual labor like a storm, bringing about a seismic shift in the way goods were manufactured. The once labor-intensive tasks were now performed with efficiency and precision by machines, leading to a surge in productivity. Coal, steel, and other energy and minerals instantly became a powerful driving force for the rapid progress of society. The demand for these resources skyrocketed, fueling the growth of industries and the expansion of cities.

Factories sprang up on the ground, transforming rural landscapes into industrial hubs. The air was filled with the sounds of machinery and the smoke from chimneys, marking a new chapter in human history. Railways were crisscrossed like steel dragons, connecting distant lands and facilitating the movement of goods and people at unprecedented speeds. The construction of these vast networks of railways was a testament to human ingenuity and ambition. Large-scale infrastructure construction changed the landscape completely, like powerful hands reshaping the earth. Canals, bridges, and roads were built, transforming transportation and trade, and enabling the flow of commerce across regions.

The industrial revolution has changed the appearance of human life and brought the human society into a new era of rapid development. Towns and cities expanded, as people flocked to urban centers in search of work and a new way of life. The skyline was dominated by the towering silhouettes of factories and the smokestacks that belched out the byproducts of industry. The streets were bustling with activity, as workers, merchants, and inventors contributed to the vibrant atmosphere of progress.

Industrialization brought about a leap of productivity, as the output of goods increased exponentially. Cities rose rapidly in this wave of progress, becoming melting pots of culture, innovation, and commerce. The population gathered like a tide, swelling the ranks of the urban workforce and creating a new social dynamic. Commercial trade showed an unprecedented prosperity, as goods from around the world were now accessible to a broader audience. The profit-driven commercial value was highlighted in this period, as businesses sought to capitalize on the new opportunities presented by industrialization.

Capital was like a galloping river, shuttling freely and maneuvering in the market, stimulating round after round of innovation and expansion. Entrepreneurs and industrialists became the new heroes of society, their names synonymous with progress and wealth. However, behind the rapid development, many thorny social problems have been exposed. The gap between the rich and the poor has intensified sharply, as the benefits of industrialization were not evenly distributed. Workers' rights and interests have been seriously infringed, with long hours, low wages, and hazardous working conditions becoming the norm for many.

Environmental pollution is deteriorating, which has become a haze over the industrial society. The smoke from factories, the waste from mines, and the pollution from new transportation networks began to take a toll on the environment. Rivers ran murky with industrial waste, and the air was thick with pollutants, leading to health problems and ecological damage. These problems are like alarm bells, prompting people to deeply reflect on the limitations of the simple pursuit of economic value. The realization grew that society needed to address these issues to ensure the well-being of its citizens and the sustainability of its environment.

In response to these challenges, there was a push to promote the value construction of social security, labor laws and regulations, environmental protection, and other institutional levels. Governments and organizations began to advocate for reforms that would protect workers and the environment, while still allowing for economic growth. Efforts were made to find a balance between development and fairness, efficiency and environmental protection, to ensure the sustainable development of society. The labor movement gained momentum, fighting for better wages, shorter hours, and safer working conditions. Environmental legislation was introduced, setting standards for emissions and waste disposal.

In this process of rapid progress, the dawn of the nineteenth century brought with it a new consciousness. The industrial revolution had not only transformed the physical landscape but also the social and political fabric of nations. It was a time of great change, where the promise of progress was weighed against the reality of its consequences. The lessons learned during this period laid the groundwork for modern society, shaping the values and policies that continue to influence our world today.

2.7 Modern Society: Collision and Integration of Globalization, Diversified Values

Since the beginning of the twentieth century, especially after the end of World War II, the world seems to have hit the fast-forward button, rapidly entering the era of globalization. The explosive progress in science and technology, coupled with the emergence and widespread adoption of the Internet, has triggered an information storm that has completely transformed the global landscape and upended traditional value systems. The cost, which can be almost negligible, and the real-time accessibility have caused knowledge to spread like a flood, at an unprecedented speed. Academic and cultural treasures, once closely guarded by a select few elites, are now displayed to the entire world. The open-source software movement has flourished, with countless passionate programmers contributing code, driven by the spirit of sharing and collaboration, creating many remarkable innovations. This movement challenges the established model of proprietary commercial software and demonstrates the immense value of shared knowledge and collective wisdom. The emergence of social media platforms has also completely changed traditional methods of personal communication, amplifying individual voices into the digital space. The virtual social domain has transcended geographical and class barriers, from cultural transmission to social movements, triggering a series of chain reactions that highlight the tremendous power of public participation and free expression in this new era. This compels us to reassess the values of the digital age, such as privacy protection and the authenticity of information, as we enjoy the fruits of technological progress while also protecting the essence of the human spirit. The wave of globalization has led to frequent collisions and integration of cultures and value systems from different countries. For example, Asian culinary culture has combined with Western fast-food culture to give birth to new food trends; African musical rhythms have merged with European classical melodies to spawn new music genres; Eastern philosophical thoughts have combined with Western scientific methods to promote the development of modern psychology. These cross-cultural exchanges and integrations have not only enriched people's lives but also facilitated the diversified development of global society. Globalization has also brought about economic interconnectivity. The rise of multinational corporations has enabled capital, goods, services, and labor to flow globally, forming an interdependent global economic network. Developing countries have gained unprecedented economic growth opportunities by participating in global supply chains, while developed countries have reduced production costs and increased competitiveness through outsourcing and foreign investment. However, this economic globalization has also brought challenges, such as job losses, lower environmental standards, and the widening gap between the rich and the poor. The field of education has also undergone dramatic changes. The emergence of online education platforms has made high-quality educational resources no longer limited to specific geographical locations, allowing anyone with internet access to take courses from the world's top universities. This democratization of education has not only raised the global education level but also nurtured a large number of talents for developing countries, promoting the modernization process of these nations. Globalization has also fostered international cooperation and dialogue. Global issues such as climate change, environmental protection, and public health require the joint efforts of all countries to address. International organizations such as the United Nations and the World Health Organization have played a significant role in coordinating national actions and formulating global policies. Through international cooperation, human society has made significant progress in addressing global challenges. However, globalization has also brought many problems and challenges, such as cultural homogenization, economic inequality, and political conflicts. How to maintain cultural diversity amidst the tide of globalization, achieve equitable economic development, and resolve international conflicts and contradictions are important issues facing countries worldwide. In the future, the direction and model of globalization will depend on the joint efforts and wisdom of the international community.

2.8 Future Outlook: Cutting-Edge Technology Challenges, Value Persistence

Standing in the forefront of the present era, human has stood in the forefront of digital technology revolution, gene editing, quantum computing, brain-computer interface cutting-edge technology breakthrough is with amazing speed, they are like a double-edged sword, open the new development door for human at the same time, also brought many unprecedented value challenges.

In the digital technology revolution, we have witnessed the rise of artificial intelligence, the proliferation of the Internet of Things, and the advent of big data analytics, all of which have transformed the way we live, work, and interact. These technologies have not only increased efficiency and connectivity but also raised concerns about privacy, security, and the displacement of jobs. As we embrace these advancements, we must navigate the delicate balance between innovation and the potential negative impacts on society.

Gene editing technology has given human beings the ability to rewrite the life code, which is expected to overcome many difficult diseases, eliminate the problems of genetic diseases from the root, and bring the light of hope to countless families. The CRISPR-Cas9 system, for instance, has revolutionized the field of genetics, allowing for precise modifications to DNA. This could lead to cures for conditions like cystic fibrosis, sickle cell anemia, and even some forms of cancer. However, followed by a series of profound ethical disputes, such as whether changing the natural genetic evolution process is legitimate, how to ensure that the technology is not abused, become a few control life, manufacturing "design baby" tools, these questions are asking the human moral bottom line, forcing us to re-examine life, the fundamental value of human nature. The prospect of "designer babies" with selected traits has sparked a global debate on the ethics of gene editing, with many calling for strict regulations to prevent a future where genetic enhancements become a privilege of the wealthy.

The emergence of quantum computing has accelerated computing to a complex scientific problem that used to take years or even decades to solve and may now be answered in an instant. Quantum computers leverage the principles of quantum mechanics to perform calculations at speeds unattainable by classical computers. This will undoubtedly greatly accelerate the science and technology research and development, password cracking, financial analysis, and many other fields of process, however, can also may cause a huge crisis of information security, traditional encryption algorithm may be useless in front of quantum computer, global network security architecture need to rebuild, personal privacy, trade secrets and national security will face severe test, which makes us have to rethink the balance between information confidentiality and sharing, and how to build with new value criteria. The potential of quantum computing to break widely-used encryption methods like RSA and ECC has prompted governments and industries to invest in quantum-resistant algorithms to protect sensitive data against future threats.

Brain-computer interface technology has broken the barrier between the human brain and the digital world, offering new possibilities for communication and control. This technology allows for direct interaction between the brain and external devices, which could be life-changing for individuals with paralysis or other disabilities. It opens up the potential for thought-controlled prosthetics, virtual reality experiences that are indistinguishable from reality, and even the ability to communicate without speaking. However, this technology also raises questions about the nature of consciousness and identity, the potential for mind control, and the implications for personal autonomy. As we continue to develop brain-computer interfaces, we must address these profound philosophical and ethical issues to ensure that the technology benefits society without compromising our humanity.

Chapter 3
The Internal Logic of the Evolution Process of Human Civilization

In the past 10 years, I have traveled all over the world, to dozens of countries and regions in Asia, Europe, Africa and Latin America, and worked with the University of Oxford, Renmin University, Peking University, Tsinghua University, The Chinese University of Hong Kong, other universities and experts and scholars to discuss the future of mankind. Discuss why conflicts, frictions and wars are frequent, and why international rules and order no longer seem to be effective. In the end, I found that the real reason and push behind it is technological progress, the Internet and the digital technology revolution.

The larger discovery is that a new civilization is forming. This is global, going beyond the countries and regions. It is based on new technologies such as the Internet, artificial intelligence, block chain and big data, as well as a more general sense of globalization. This new civilization has brought great changes and challenges, which is the answer to the century question that I have explored for many years! This is the darkness before the arrival of a new civilization, and although the darkness will pass, it is painful. We must accept this new reality, because the pain may not be short-term, but a long-term trend.

However, the essential reason behind the chaos and crisis in the world is not the development of this new civilization itself, but the many problems and challenges that arise in the process. One of the main problems is the shift of values and rules. In this new civilization, changes in values and rules are more important than technological and economic changes. We have seen the collision and conflict of many different cultures and values in this new era of globalization. These cultural differences and different values often cause many conflicts and tensions.

In addition, as technology evolves, the social structure also changes. In a new civilization, a new structure and organizational form of the international society will appear accordingly. This will lead to the loss of control and authority of many traditional organizations and institutions, thus to instability and contradictions in the international community. We therefore need to develop new rules and values to

L. Yan, *Value-Based Civilization*, Understanding China,
https://doi.org/10.1007/978-981-96-9490-7_3

create a sustainable, peaceful and stable global order. This requires cooperation and collaboration between the international community and international organizations.

The development of digital technology is undergoing an unprecedented revolution, especially since the emergence of block chain in recent years has injected greater impetus into the revolution. New digital technologies are changing our lives. They not only have a profound impact on the progress and development of human society, but also reconstruct the world pattern and international relations. However, due to the rapid development of digital technology and the new technological revolution, our international community, especially the major powers that once dominated the world situation, seems unable to fully adapt to these changes. In the past, the international rules and orders we know are based on industrial civilization and information civilization, and these rules and orders have been unable to adapt to the world development trend under the new technological revolution. As a result, the international community has indications, including war, friction and withdrawal. These problems have led to the turbulence and changes in the rules, order, pattern and power of the world, and made us feel the pain of the emergence of a new civilization.

As an important milestone in the development process of mankind, the digital technology revolution is bringing us new opportunities and challenges. In particular, with the emergence of block chain technology, this technology can not only accelerate the digital process, provide people with a safer and more convenient way to trade, but also bring a fair and transparent governance environment, and expand the choice and autonomy of participants. It will also be an important opportunity to rebuild world rules and order.

So what are the boundaries and conditions for the emergence of a new human civilization? What are the rules to be followed, and with these questions, I continue my research. Hard work pays off. I finally discovered the origin and essence of human civilization, as well as the core and driving force of human civilization. The core of human civilization is "value", while the driving force of human civilization is "technology". This is a major discovery, and I have reviewed a lot of literature and found that this is the first time in human history that human civilization has been studied from this perspective.

In the long and magnificent development of human civilization, many elements interweave and collide with each other, converging into a mighty force, like a powerful engine, driving the great ship of history rolling forward, spanning the long years, and engraved countless glory and vicissitudes. When we with deep and keen eyes, through the fog of history, explore the evolution of internal logic, will be discovered, technology, value and negative entropy like three key master key, bear the driving force, the kernel and password, they continue to inject vigor and vitality for human civilization, accurately guide the direction of the forward, cleverly unlock development on the way.

3.1 Technology: A Strong Driver of the Evolution of Human Civilization

In the long picture of human history, technology has always played a pivotal role. It is like a powerful force, pushing the society forward and shaping the colorful face of civilization. From the birth of simple but significant simple tools in ancient times to the rapid development of modern amazing science and technology, technology, with its unique charm and great influence, has profoundly changed every corner of human society.

The origin of technology can be traced back to the dawn of human civilization. At that time, technology quietly appeared in the form of primitive tools, like the first ray of new bud in the spring, bringing new hope and possibilities to human beings. By observing nature and imitating living creatures, ancient humans created simple tools such as stone tools and wood tools. These seemingly rough inventions contain human wisdom and creativity, greatly improving the survival ability of ancient humans, so that they can better hunt, gather food and resist the threat of nature.

In the Paleolithic age, the early human beings with the initial observation and unremitting exploration of nature, learned to polish the stone tools, the key skill to change the fate, carefully made such as simple and practical cutter, scraper and other tools. Although these instruments are unremarkable in the modern view, at that time, they gave humans the magical power to initially transform nature. With these tools, humans can hunt and gather more efficiently, and obtain essential food and precious resources from the vast expanse of nature. This not only greatly improves the survival probability of human beings, but also promotes the gradual change of the role of human beings, from the weak who passively adapted to the natural environment in the past to the warriors who took the initiative and explored the mysteries of nature, and resolutely opened the great preface to the evolution of civilization.

As time goes by, the Neolithic Age comes lightly. Agricultural technology is like a seed of hope breaking from the soil, taking root and sprouting in the soil of human history. Humans began to try to tame animals and plants, and gradually mastered the core skills of growing grain and raising livestock. This revolutionary change of great significance has changed the life mode of human society dramatically, from a steady nomadic life to a relatively stable and stable settled life. Villages mushroomed after a spring rain on the earth, the population began to gather, social division of labor also began to appear. Some focus on the fields and work hard to provide a stable source of food for the community; others devote themselves to crafts, awe, and religious sacrifices. In this way, it laid a solid foundation for the further prosperity of civilization.

Later, the Bronze Age and the Iron Age came one after another, and the wide application of metal tools was like a storm sweeping the earth, bringing earth-shaking changes to the human society. The emergence of metal farm tools increased the efficiency of agricultural production as a rocket, increased the output of agricultural products, feeding the growing population; the birth of metal weapons completely changed the pattern of war, promoted the formation and expansion of countries, and rewritten the map of human history. In this magnificent development process, words, as a technological invention with epoch-making significance, are like a bright star, cutting through the dark night sky of knowledge dissemination. It breaks the limitation of information transmission from word of mouth, enables knowledge and experience to be systematically recorded and inherited, creates unique conditions for the vigorous development and prosperity of science, literature, art and many other fields, and opens a new era of the light of human wisdom.

The industrial revolution is undoubtedly a highlight moment in the evolution of technology-driven civilization, like a shocking tsunami of science and technology, sweeping the world with a devastating momentum. The emergence of steam power and widely used, like a dose of super energy stimulant into the traditional production mode, opened a new era of mechanized production. Factories are springing up in every corner of the city, traditional handicrafts are quickly replaced by the roar of machines, and urbanization accelerates like a high-speed train. In the field of transportation, the birth of new means of transportation has happened. Trade and cultural exchanges around the world have become unprecedented convenient and efficient, and the earth seems to be instantly compressed into a "global village". Since modern times, the evolution of technology has been accelerated. The invention of the steam engine opened the industrial era, like a storm sweeping through the world. Machine production replaced manual labor, and production efficiency was unprecedented. Factories are springing up and a large number of goods are produced to meet peoples growing material needs. The development of railways, ships and other means of transportation has facilitated the circulation of goods and promoted the prosperity of international trade. The wide application of electricity has brought human society into the age of electrification, as in the dark Lamps to illuminate life. Light and power continuously flow into thousands of households, and peoples living conditions have been greatly improved. The invention of electric light makes the night no longer dark, and the emergence of electric motor allows all kinds of mechanical equipment to operate. Electric power has become a powerful driving force for social development and brought unprecedented convenience to human beings. Like a pair of powerful wings, to further expand the boundary of human activities, so that the human society to stride into the fast lane of modernization, toward a higher level of civilization rapidly stride.

Today, in the tide of the information age, the rapid development of information technology is like a strong east wind, but also pushes human civilization to a new peak height. The popularity of the Internet is like an invisible net, breaking the heavy constraints and boundaries of time and space, so that information can spread to every corner of the world in an instant. People can know the world affairs without leaving home, and the acquisition of knowledge is as easy as a bag. The booming rise of e-commerce, like a revolution in the commercial field, has completely changed the traditional business model, greatly improved the efficiency of economic operation, and made shopping more convenient and efficient. The continuous emergence of cutting-edge technologies such as artificial intelligence, big data and biotechnology, like a group of energetic innovation pioneers, is breeding a new round of scientific and technological revolution and industrial revolution, drawing a grand blueprint full of infinite possibilities for the future development of human civilization, making people full of expectations and expectations for the future.

Technological innovation has also promoted the birth and development of emerging industries. High-tech industries such as information technology, biotechnology and new energy have risen rapidly and become new engines of economic growth. The development of these industries has created a large number of employment opportunities and promoted the optimization and upgrading of the industrial structure. Traditional industries are constantly transformed and upgraded driven by technology to improve product quality and added value. For example, the manufacturing industry introduces intelligent manufacturing technology to realize the intelligence and digitalization of the production process, and improve the production efficiency and product quality.

In society, technology also plays an important role. It has changed peoples communication style and social habits, making the communication between people more convenient and efficient. Today, people can keep in touch with others through social media and instant messaging tools, and share their lives and experiences. This new way of networking breaks the limits of time and space, making it easy for people to make friends and expand their social circles.

Technology also promotes the communication and integration between different cultures. The popularity of the Internet has made people understand the cultural customs, art works and ideas of different countries and regions. Through online videos, music platforms and other channels, people can enjoy excellent cultural works from all over the world to enrich the spiritual world. The communication and collision between different cultures stimulate peoples creativity and innovative spirit, and promote the progress of human civilization.

Technology has driven the change and innovation in education. New education models, such as online education and distance education, will enable more people to enjoy high-quality education resources and achieve equitable and universal education. Students learn knowledge anytime and anywhere through the network platform, and interact with teachers and students. The application of educational technology improves the efficiency and quality of teaching, and makes learning more interesting and efficient.

In terms of culture, the influence of technology cannot be ignored. It has enriched peoples cultural and entertainment life, and the popularization and development of film, TV, music and other art forms cannot be separated from the support of technology. With the help of advanced audio and video technology and special effects means, the artists create more shocking works, so that the audience is immersed in the ocean of art and feel the charm and fun.

From silent to sound, from black and white to color, from 2D to 3D, every technological advance brings a new visual experience. The application of special effects technology makes the fantasy scenes in the film more realistic, making the audience as if they are there. The development of TV allows people to enjoy a variety of wonderful programs at home and enrich their spare time life. Music is also constantly innovated by technology. The emergence of digital music and online music platforms makes it easier for people to appreciate and share music works.

Technology has also promoted the protection and inheritance of cultural heritage. Through digital technology and virtual reality technology and other means, the precious cultural heritage can be preserved in the form of digital, and displayed and spread through the Internet and other platforms, so that more people can understand and appreciate these treasures. Digital technology scans and models cultural relics with high precision to preserve detailed information of cultural relics. Virtual reality technology enables people to feel the charm of cultural heritage personally, and enhance the awareness of protection.

In life, technology is everywhere. Smart home system makes life more intelligent and convenient. Through smart phones or smart speakers, people remotely control home electrical equipment to adjust lights, temperature, humidity, etc. The emergence of smart home appliances makes housework easier and improves the quality of life.

Advanced medical technology provides more efficient and accurate treatment means. The application of medical imaging technology, genetic testing technology and minimally invasive surgical technology enables doctors to diagnose diseases more accurately and formulate scientific and reasonable treatment plans. The development of telemedicine technology enables patients to enjoy high-quality medical services at home and ease the shortage of medical resources.

The development of environmental protection technology allows us to better protect the earth while enjoying modern life. The application of new energy technology reduces the dependence on traditional fossil energy and reduces carbon emissions. Progress in sewage treatment technology and waste treatment technology will make the environment cleaner. The promotion of green building technology makes buildings more energy-efficient and environmentally friendly, and provides people with a comfortable living and working environment.

In social governance, the application of technology has improved the government management efficiency and public service level. With the help of big data, cloud computing and other advanced technologies, the government can more accurately understand the needs and demands of the people. By analyzing a large amount of data, the government finds the hot spots and difficulties of social problems, and formulated scientific and reasonable policies and measures.

Through the government service platform and other channels, the government will make it easier for the government to communicate and interact with the public and respond to social concerns and problems in a timely manner. People can handle government affairs anytime and anywhere through mobile phones, computers and other equipment. The transparency and credibility of the government have been enhanced, and peoples satisfaction with and trust in government work have increased.

The technology-supported governance model has improved the efficiency of government work and enhanced the accuracy and effectiveness of social governance. For example, in urban management, intelligent transportation system, intelligent security system and other technical means to achieve real-time monitoring and control of traffic flow, improve traffic efficiency, and ensure urban safety. In emergency management, big data technology helps the government to quickly respond to emergencies, rationally allocate resources, and improve the efficiency of emergency rescue.

Technology has contributed a lot in promoting social progress. Technological progress promotes social equality and justice. The development of information technology such as the Internet has broken down information barriers and geographical restrictions, and enabled more people to obtain rich knowledge and information resources. Wherever you are, with an Internet connection, you can enjoy global educational resources, cultural resources and business opportunities.

The popularization and application of new energy technologies have reduced the dependence on traditional energy sources, reduced environmental pollution and resource waste, and promoted the sustainable development of the society. The use of renewable energy sources such as solar, wind and water energy not only reduces carbon emissions, but also provides power supply to poor areas and promotes economic development. Technological advances allow us to share the earth's resources more fairly and achieve common development.

Technology has stimulated the spirit of social innovation and change. The rise of e-commerce and the prevalence of sharing economy cannot be separated from the support and promotion of technology. E-commerce has changed the way people shop, allowing consumers to buy global goods just at home. The sharing economy makes full use of idle resources and improves the efficiency of resource utilization.

These emerging economic forms have changed people's consumption style and living habits, bringing more employment opportunities and entrepreneurial space to the society. People carry out entrepreneurial projects through the Internet platform to realize their life value. The innovation and application of technology promote sustained economic prosperity and development and inject new vitality into social progress.

However, we should also look at the challenges and problems that technology may bring in social governance and social progress. With the rapid development of digital technology and the expansion of its application scope, information security, privacy protection and other issues have become increasingly prominent. In the Internet age, the disclosure of personal information has become a serious problem. Hacker attacks, network fraud and other incidents occur from time to time, threatening people's property security and personal privacy.

The rise of technologies such as artificial intelligence has had a profound impact on the employment structure and social ethics. The application of artificial intelligence may lead to the disappearance of some traditional occupations, such as manufacturing workers and customer service workers. At the same time, the development of artificial intelligence has brought some ethical problems, such as whether the decision of artificial intelligence is fair and whether artificial intelligence will surpass human beings.

In short, technology plays a key role in the development of human civilization. From the invention of ancient tools to the rapid development of modern science and technology, technology has profoundly changed all aspects of human society. In the future, technology development will pay more attention to sustainability and humanization. The development of new energy technologies and environmental protection technologies will provide us with a cleaner and more sustainable energy and living environment. The development of artificial intelligence, biotechnology and other fields will bring more benefits to human beings. For example, the progress of medical technology will make people live a more healthy and long life, and the application of intelligent robots will provide people with more convenient services. Technology is a powerful driving force for the development of human civilization. We should give full play to the advantages of technology, actively respond to the challenges brought about by technology, and so that technological progress can better serve the development of human society.

3.2 Value: The Core of the Evolution of Human Civilization

In the expansive domain of value philosophy, the concept of value transcends mere material worth, delving into the philosophical discourse that examines the intricate relationship between an object and its ability to meet the needs and desires of a subject. This philosophical exploration encompasses not only the efficacy and benefit that an object's attributes and functions provide to a subject but also the broader impact on the subject's well-being and fulfillment. To achieve a profound comprehension of the evolution of civilization, I have immersed myself in an extensive array of literature, scholarly articles, historical documents, and anthropological studies. Through this exhaustive research, it has become evident for the first time that value is not merely a peripheral aspect but indeed lies at the very heart of human civilization. It serves as the foundational cornerstone upon which the grand edifice of human societal evolution is built; it also acts as a guiding light, illuminating the path for humanity as it traverses the vast and intricate landscape of historical progression.

The origins of value can be probed on multiple levels, each revealing a different facet of its significance. From the perspective of human needs, especially during the nascent stages of societal development, survival is of paramount importance. In these early times, the fulfillment of basic physiological requirements—such as sustenance, shelter, and safety—constitutes a pivotal source of early value. During this era, value is primarily concerned with satisfying these fundamental physiological demands.

For instance, consider the role of food in early human societies. Items such as food procured through hunting, gathering, and later through rudimentary forms of agriculture, hold immense value, as they are directly linked to the survival and sustenance of individuals and communities. The successful hunt that brings back meat or the harvest that yields grains and vegetables is not just a matter of nourishment; it is a triumph over the harshness of nature, a celebration of life, and a source of communal bonding. Similarly, the discovery and utilization of caves or the construction of simple shelters offer refuge from the elements, protection from predators, and a sense of security, thus becoming highly valued assets in the early human experience.

Moreover, the development of tools and the innovation of basic technologies, such as the creation of fire for warmth and cooking, or the crafting of rudimentary clothing from animal hides, further illustrates the manifestation of value in early human civilization. These advancements not only enhanced the quality of life but also represented significant leaps in human capability, each step forward marked by the creation of new values that propelled humanity towards greater heights.

As societies evolved, so too did the concept of value. Beyond the mere satisfaction of physiological needs, value began to encompass social, cultural, and spiritual dimensions. The establishment of social norms, the creation of art and symbols, and the development of language and communication all contributed to a richer, more complex tapestry of values that shaped human civilization. These values, in turn, influenced the formation of communities, the establishment of governance structures, and the pursuit of knowledge and wisdom.

In conclusion, the concept of value in the realm of philosophy is not static; it is dynamic and multifaceted, reflecting the ever-changing needs and aspirations of humanity. By exploring the historical and philosophical dimensions of value, we gain a deeper appreciation for the forces that have shaped human civilization and continue to influence our journey through the vast expanse of historical progression.

As human intelligence gradually unfolds, the pursuit of knowledge, the inheritance and creation of spiritual culture, and the quest for understanding the mysteries of the universe become important value origins that drive our species forward. The manufacture of tools, from the earliest stone axes to the sophisticated machinery of today, becomes a significant sign of civilization progress. A carefully polished stone axe, a sharp bone spear, not only improve hunting and gathering efficiency, but also represent humanity's initial attempts to harness and transform nature to better serve our needs. Their value far exceeds the material itself, embodying humanity's steps towards civilization and the mastery of our environment.

As social organization becomes more complex and division of labor begins, the connotation of value also becomes increasingly rich and diverse. In the economic field, value origins include the improvement of production efficiency, the acquisition of commercial benefits, and the optimization of resource allocation. These factors contribute to the wealth and prosperity of societies, as well as the well-being of individuals within them. At the social level, fairness, justice, harmony, and equality are also value origins, promoting the construction of social security systems, labor laws, and other institutions that protect the rights and interests of citizens. These values are essential for maintaining social order and ensuring that progress is inclusive and equitable.

In agricultural society, fertile land becomes the cornerstone of civilization, carrying people's hopes and expectations for a bountiful harvest. It is not only a symbol of wealth but also the foundation of survival, as it provides the necessary resources for sustenance. This reflects the value origin of improving production efficiency and ensuring food security, which are critical for the sustenance of growing populations. Craftsmen, with their exquisite skills, make beautiful pottery and textiles. These creations not only meet the practical needs of daily life but also become carriers of cultural heritage and artistic aesthetic expression. They embody the value of human creativity and the importance of preserving cultural identity through tangible artifacts.

The invention of writing marks a monumental revolution in value. It breaks the limitations of knowledge being passed down only by word of mouth, allowing human experience and thoughts to be inherited across time and space. This innovation paves the way for the recording of history, the codification of laws, and the dissemination of knowledge, which in turn fosters the vigorous development of philosophy, literature, science, and many other fields. The written word becomes a vessel for ideas, enabling the accumulation of wisdom and the advancement of civilization in ways that oral traditions could never achieve alone. It is through this written legacy that we continue to build upon the achievements of our ancestors and propel humanity towards an ever more enlightened future.

As humanity stepped into the industrial age, the thunderous clamor of machinery erupted, shattering the serene ambiance of age-old production techniques. The relentless hum of engines and the clanking of gears became the new soundtrack of progress. Coal, with its dark, sooty hands, and steel, with its gleaming, resilient strength, emerged as the cornerstones of a new era, propelling society forward at an unprecedented pace. These energy minerals, once obscure and unassuming, now stood as symbols of power and advancement, fueling the fires of innovation and industrial might.

The construction of large-scale infrastructure, such as sprawling factories and expansive railway networks, marked a dramatic transformation in the landscape of human existence. Factories, with their towering smokestacks piercing the sky, became the new cathedrals of industry, where the symphony of production played out day and night. Railways, those iron veins, crisscrossed continents, knitting together distant lands and cultures, enabling the rapid movement of goods, people, and ideas. These monumental projects not only reshaped the physical world but also served as a catalyst for economic growth and societal change.

The industrialization value they represented spread like wildfire across the globe, igniting a revolution in productivity that had never been seen before. This surge in productivity development was akin to a tidal wave, sweeping away the old ways and ushering in a new era of efficiency and output. The transformation was so profound that it reshaped social structures and the very fabric of people's value perceptions. Traditional roles and hierarchies were upended, giving way to new classes and power dynamics.

Cities, once small and isolated, rose rapidly like phoenixes from the ashes of the past, becoming bustling metropolises teeming with life and opportunity. Populations flocked to these urban centers, drawn by the promise of work and a better life, leading to unprecedented population density and urbanization. Commercial trade flourished as never before, with markets expanding beyond local bazaars to global exchanges. The pursuit of profit-oriented commercial value became not just prominent but the dominant ethos of the age, driving individuals and corporations alike.

Capital, the lifeblood of the economy, flowed freely in the market, stimulating innovation and expansion at an exponential rate. New technologies and industries sprang up like mushrooms after rain, each promising greater wealth and power. However, this unbridled pursuit of economic value also brought about a host of social issues. The chasm between the rich and the poor widened, creating a landscape of stark contrasts and inequality. Labor rights were often trampled upon, with workers toiling in unsafe conditions for meager wages, sparking a growing awareness and pushback against these injustices.

This dichotomy between progress and its consequences prompted a profound reflection on the limitations of a single-minded focus on economic value. People began to question whether the relentless pursuit of profit should come at the cost of social well-being and justice. In response to these concerns, the construction of value at the institutional level began to take shape. Social security systems were established, providing a safety net for the vulnerable and ensuring a basic standard of living. Labor laws were enacted, protecting workers' rights and setting standards for fair wages and safe working conditions. These institutional frameworks were built upon value origins such as social fairness, justice, and the well-being of the populace, marking a shift towards a more balanced and humane approach to progress.

In the information age, the birth of the internet was like an information storm that completely overturned the traditional value system. Information dissemination is nearly zero-cost and instantaneous, allowing knowledge to spread at an unprecedented speed. Academic and cultural resources that were once monopolized by a few elites are now accessible to the masses. The open-source software movement has flourished, with countless programmers voluntarily contributing code, creating many outstanding products based on the concepts of sharing and collaboration, challenging the traditional closed-profit model of commercial software and showcasing the tremendous value of knowledge sharing and the aggregation of group wisdom. This reflects the expansion of value connotations and the change of value origins in the new era.

The internet has revolutionized the way we access and distribute information, effectively democratizing knowledge. In the past, obtaining information required physical access to libraries, archives, or other institutions that held the data. Now, with just a few clicks, one can access a wealth of information from anywhere in the world. This has led to a more informed and educated public, as well as a more level playing field in terms of who can contribute to the global conversation. The open-source software movement is a prime example of this shift. Projects like Linux, Apache, and Mozilla Firefox have been developed by communities of programmers who share their work freely, allowing for the creation of high-quality software that competes with and often surpasses proprietary alternatives. This collaborative approach not only reduces costs for consumers but also fosters innovation and learning within the tech community.

Social media platforms have changed the mode of interpersonal communication. Individual voices are amplified infinitely, virtual social circles cross geographical and class boundaries, It causes a series of chain reactions from cultural communication to social movements, which makes us see the power of public participation and freedom of expression in the new era, and also urges people to re-examine the accompanying value issues such as privacy protection and information authenticity, which is also the embodiment of the existence and evolution of value in the information age. The rise of social media has fundamentally altered how we interact with one another. Platforms like Facebook, Twitter, and Instagram have given individuals the power to broadcast their thoughts and experiences to a global audience. This has led to a more connected world where people can share their lives, ideas, and causes with others who may have never had the opportunity to connect otherwise. However, this new form of communication also brings challenges, such as the need to navigate issues of privacy and the spread of misinformation. As we grapple with these issues, we are also redefining the values that underpin our interactions in the digital age.

Looking back at history, different civilizations, based on unique geographical environments and development paths, have given birth to various core values, collectively painting the colorful tapestry of human civilization. In the ancient Chinese civilization of the East, the "benevolence, righteousness, propriety, wisdom, and trust" advocated by Confucianism Penetrating all levels of society. These values have shaped the way Chinese society has functioned for millennia, influencing everything from family structures to governance. Confucianism's emphasis on moral virtue and social harmony has been a cornerstone of East Asian culture, and its principles continue to resonate in modern society. As we move further into the information age, we are seeing how these traditional values interact with new digital realities, creating a complex interplay between the old and the new.

In contrast, the ancient Greek civilization, the source of Western civilization, is characterized by its relentless pursuit of rationality as its value beacon. From Thales' exploration of the origin of all things to kick off the tradition of rational speculation, to Socrates' inspiring public discourse on the streets about the truth and essence of virtue, to Aristotle's construction of a vast knowledge system encompassing philosophy, natural science, political science, and many other fields, using the light of reason to dispel the darkness of ignorance, laying the foundation for the development of Western academic thought. This spirit of rationality permeates city-state politics, where citizens engage in public debates and vote on public affairs, with the rudiments of democratic systems emerging, reflecting the value of reason in the political realm; reflected in artistic creation, ancient Greek sculptures showcase a reverence for human beauty and rational order with precise proportions and dynamic poses, and the Parthenon, with its rigorous geometric structure and harmonious proportions, demonstrates the pinnacle of rational aesthetics, setting a benchmark for Western art in later times, showing the value of reason in the fields of culture and art.

In the contemporary era swept by the wave of globalization, diverse value systems are crashing into one another and melding at an unprecedented rate. The world has become a melting pot where ideas, traditions, and philosophies from different cultures are constantly interacting and influencing each other. This phenomenon is particularly evident in the realm of global enterprises, which have extensive business networks that span the globe. These corporations are tasked with managing a workforce that is rich in cultural diversity, bringing together individuals from various backgrounds, each with their unique perspectives and ways of working.

It is incumbent upon these global enterprises to honor the Western ethos of individual performance incentives to foster employee innovation. This means recognizing and rewarding individual contributions and achievements, which can motivate employees to push the boundaries of their creativity and productivity. At the same time, these companies must also embrace the Eastern camaraderie that binds collective efforts towards problem-solving. This involves promoting teamwork and collaboration, where the success of the group is valued over individual glory, and where employees work together to achieve common goals.

This constitutes a tangible fusion of disparate value systems within the economic sphere. The challenge for global enterprises is to balance these contrasting values, creating a harmonious work environment where both individual creativity and collective effort are valued and nurtured. This delicate balance can lead to a more innovative and productive workforce, capable of tackling the complex challenges of the modern business world.

Similarly, in international scientific research collaborations, scholars from varied educational pedigrees and intellectual traditions converge. The Western predilection for empirical evidence and the challenging of established norms in scientific pursuit is invigorated by the Eastern methodical scholarship and its meticulous progression. Western scientists often emphasize hypothesis testing and the rigorous application of the scientific method, pushing the boundaries of what is known and questioning long-held beliefs. This approach is complemented by the Eastern methodical scholarship, which values a step-by-step approach to research, building upon existing knowledge with patience and precision.

This exemplifies the interplay and amalgamation of values in the scientific realm. The result of this fusion is often groundbreaking innovations that transcend cultural boundaries. For instance, the development of cutting-edge medical technologies or the exploration of space benefits from the integration of different scientific philosophies and methodologies.

In the cultural exchange domain, Hollywood films adeptly weave in Chinese martial arts and Eastern philosophies to captivate a worldwide audience. Movies like "Crouching Tiger, Hidden Dragon" and "Kung Fu Panda" have successfully integrated elements of Eastern culture into mainstream cinema, showcasing the beauty and depth of Eastern storytelling and visual arts. These films not only entertain but also educate global audiences about the richness of Eastern culture.

While Western cinema has been incorporating Eastern elements, Chinese web novels are exporting the nation's fantastical imagination and traditional ethos, igniting a craze in the global cultural marketplace. These novels, with their unique blend of Eastern mythology, history, and philosophy, have gained a massive following outside of China. They offer readers a glimpse into a world where ancient wisdom and modern storytelling converge, creating a new form of global entertainment.

This vividly illustrates the confluence of cross-cultural values. As these cultural products gain popularity worldwide, they not only entertain but also foster a greater understanding and appreciation of different cultures. This cultural exchange helps to break down barriers and build bridges of mutual respect and understanding between people from diverse backgrounds.

Looking to the future, humanity stands at the forefront of a technological revolution. Breakthroughs in frontier technologies such as gene editing, quantum computing, and brain-computer interfaces will once again rewrite the value map. Gene editing holds the promise of conquering intractable diseases but raises ethical controversies over "designer babies," forcing humanity to redefine values such as the dignity of life and natural selection, involving new considerations of life values;

quantum computing's powerful computing power will overturn existing encryption systems and accelerate scientific research processes, bringing about changes in information security and knowledge acquisition methods, triggering a re-examination of the value of information and the value of knowledge; if brain-computer interfaces achieve deep integration between humans and machines, human cognition, emotions, and machine intelligence will interact, challenging traditional value boundaries such as self-awareness and free will, prompting humanity to deeply reflect on its own values. Faced with these unknowns, humanity needs to learn from history, base itself on the common well-being of all humanity, and draw nourishment from the wisdom of ancient civilizations and modern development experiences to build an inclusive and sustainable value framework. Because value, as the core of human civilization, is our anchor to navigate through the thick fog and face the turbulent waves. Whether it is the splendid achievements of the past, the vibrant development of the present, or the infinite possibilities of the future, all require the guidance and support of value. Only by tightly grasping this core can humanity find direction in the ever-changing tides of the era, gather strength, continue the immortal legend of civilization, and forge ahead towards a brighter, more harmonious, and prosperous.

3.3 Negative Entropy: The Mysterious Code for the Evolution of Human Civilization

"Entropy", originally a professional concept born in the field of thermodynamics, is used to accurately describe the disorder degree of the system, like a chaotic fog, enveloping the development trend of the system. And opposite "negative entropy", is like a dawn through the fog, means against disorder, towards orderly force, it is like a hidden behind the evolution of human civilization, mysterious password lock, quietly reveals the human society from confusion to orderly clear, from low humble to advanced delicate inner core mechanism, we interpret the development of human civilization provides a new perspective.

From a unique perspective of biology, life itself is an amazing negative entropy phenomenon, like an orderly island, standing tenaciously in the cosmic ocean of entropy growth. Biology is like an exquisite engineer, by absorbing external material and energy, continuously absorb nutrients like sponge absorbing water, and then methodicdischarge disorderly waste, so as to maintain its highly orderly internal structure, so as to realize the magical process of growth, development and reproduction. As the highest level form of life on earth, human beings play this negative entropy characteristic to the extreme, like a wise master. Proprimitive anthropology learned to make fire, a seemingly simple behavior, but actually contains profound negative entropy wisdom. Cooking food with fire turns raw food which is difficult to digest and may carry germs into delicious and safe cooked food; combat the disordered cold and darkness of natural environment and bring warmth and light to human life. This pioneering move opened the first use of negative entropy, like the first light of human civilization.

At the social level, the establishment of the system, like a solid fortress, is an important means for human beings to achieve negative entropy. Early tribal society, through the established custom, taboo and simple form of social organization, like an invisible safety net, regulate the behavior of members, avoid internal conflict and chaos, ensure the stable operation of the tribe, as anchored in the raging river, let the boat of tribe smoothly. With the development of the society, the emergence of the country is like a towering lighthouse, bringing a more complex and perfect system. The legal system is like a fair sword, clarifies the rights and obligations, maintains social fairness and justice, enables the social order; the political system is like a precise power structure, establishes a reasonable framework of power operation, guarantees the order of social decision-making and governance, ensures that power is not abused; the economic system is like a pair of dexterous hands, standardizes the whole process of production, distribution and exchange, promotes the rational allocation of resources, and enables economic activities. These systems are like precise bite gears, cooperating with each other, promoting the orderly operation of the huge machine of society, like a harmonious symphony performance, fighting against the disorder and chaos that may occur in the natural state The development of human society is protected.

Cultural inheritance is also a vivid embodiment of negative entropy, like a flowing cultural river, flowing with the crystallization of human wisdom. Through language, writing, art and other colorful forms, like a faithful messenger, the ancestors accumulated knowledge, experience and skills from generation to generation, so that each generation can stand on the shoulders of predecessors, inherit the wisdom of predecessors, rather than starting from scratch, groping forward in the dark. The accumulation of this culture is like an increasing knowledge pyramid, enabling human civilization to evolve, avoiding the dilemma of stagnation or regression due to the temporary nature of individual life, and like injecting continuous power into the development of human civilization, and making human society move forward steadily in the long river of history. Part of ancient classics, carrying the history of ancient works of art, contains the originality of traditional craftsmanship, like time capsule, bearing the essence of human wisdom, they through the tunnel, passing the orderly information and energy, power the human society towards a higher level of orderly development, constantly climb the peak of civilization.

Scientific and technological innovation is injected with a powerful negative entropy force for human beings, like a magical magician, constantly create miracles. Every technological breakthrough is like a key, means that human deeper understanding of nature and accurate control, can more effective use of natural resources, like a budget housekeeper, improve the production efficiency, reduce the waste of resources and environmental damage, so that the human society in the process of development to keep relatively low entropy, like a smooth sailing in the turbulent sea. For example, the vigorous development of clean energy technology is like a green revolution, helping to reduce the dependence on traditional fossil energy, reduce environmental pollution, realize efficient and sustainable energy utilization, and make a green protective clothing; Intelligent manufacturing technology is like an intelligent craftsman, can optimize the production process, accurately allocate resources, improve product quality and production efficiency, promote industrial upgrading and economic development, make human production and life more intelligent and convenient.

In the face of global challenges, international cooperation also plays a key role in negative entropy, like a pair of big hands, across national borders and gathering strength. Climate change, infectious disease pandemics, terrorism and other issues are like a group of fierce monsters, crossing national borders, threatening the survival and development of all mankind, and leading the world into disorder and panic. Countries through working hand in hand, like a coalition, sharing information, technology and resources, develop common strategies, like a group of wisdom, common plan, can form a powerful force, effectively resolve the crisis, maintain global stability and order, like a strong wall, resist the invasion of the outside world. This coordinated action beyond national interests reflects the wisdom and determination of mankind to pursue negative entropy and fight against disorder in a larger scope. It is like a banner of unity and demonstrates the common destiny and responsibility of mankind.

To sum up, negative entropy, as the mysterious code of the evolution of human civilization, is like an invisible line, running through all aspects of the development of human society. It prompted human energy and information from the outside world, like a greedy knowledge, through their own efforts and creation, like a industrious craftsmen, build orderly social structure, inheritance and orderly cultural knowledge, orderly development system of science and technology, in the entropy of the universe opened up a evolving, to the unique path of civilization, like a brave pioneer, develop infinite possibilities for the future of mankind.

Technology, value and negative entropy are like closely interlocking gears, which are interrelated and complement each other, and together constitute the internal logic of the evolution of human civilization. The progress of technology is like a solid bridge, providing material support for the realization of value, so that the ideal pursuit of human beings can take root, like the seeds of hope; value is like a beacon, pointing out the direction for technology development, ensure technology serves human welfare and the progress of civilization, make the boat of technology does not deviate from course; and negative entropy as a loyal guard, guarantee the human society under the dual drive of technology and value, always maintain orderly development, toward a higher level of civilization, like a firm navigator, leading human to the other side of civilization. In the future journey, a deep understanding and grasp of this internal logic will help mankind to cope with various challenges more calmly, like a wise warrior, to continue to write a more brilliant chapter of human civilization, like an outstanding poet, to write a magnificent epic of the future of mankind.

Chapter 4
The Human Civilization Function

4.1 Human Civilization Function Model (HCF Model)

When we embark on a journey to explore the depths of human civilization, we can construct a comprehensive and functional model that serves to elucidate its intricate and sophisticated operational mechanisms. This model acts as a blueprint, offering insights into the inner workings of societal progress. It is akin to a detailed map that charts the course of human advancement, highlighting the pivotal role of technology and the evolution of value systems.

To illustrate, let's consider a chart that provides an enlightening perspective on the symbiotic relationship between technological innovation and the shifting paradigms of value. This chart is not merely a static image; it is a dynamic tool that offers an intuitive and systematic viewpoint, allowing us to trace the intricate web of connections between the two. It acts as a compass, guiding us through the vast landscape of human history, pinpointing the landmarks of significant technological milestones and the corresponding shifts in value cognition.

The human civilization function model is an intellectual instrument designed to decode the internal logic that underpins the development of human societies. It posits technology and value cognition as the core variables that drive the engine of civilization. By examining the interplay between these two elements, the model reveals the unique characteristics of different stages of civilization. It provides a systematic framework that enables us to dissect the anatomy of human society, analyzing its growth and transformation over time.

Within this framework, technology emerges as an independent variable, a catalyst that propels the forward momentum of civilization. The advent of new technologies at various historical junctures has precipitated comprehensive social change. For instance, the invention of the wheel revolutionized transportation and trade, while the printing press democratized knowledge and information. Each technological leap has not only transformed the way humans interact with their environment but also reshaped societal structures and economic systems.

L. Yan, *Value-Based Civilization*, Understanding China,
https://doi.org/10.1007/978-981-96-9490-7_4

Conversely, value cognition serves as the dependent variable, dynamically adjusting to reflect the value demands and evaluation criteria of different periods. As societies evolve, so do the values that underpin them. The shift from agrarian to industrial societies, for example, saw a transformation in the value placed on labor, from manual to mechanized work. This transition was accompanied by a reevaluation of the worth of individual contributions and the emergence of new social norms and ethical considerations.

At the heart of this model lies the concept of value, which remains a constant throughout the ages. It represents the enduring commitment of humanity to universal values such as the pursuit of a better life, fairness, and justice. These values are the guiding stars that have illuminated the path of human progress, inspiring innovations and shaping the moral compass of societies. They are the principles that have driven movements for social change and have been the bedrock upon which civilizations have been built and sustained.

In conclusion, the human civilization function model, enriched by the enlightening chart, offers a multifaceted understanding of the complex interplay between technology and value cognition. It is a testament to the dynamic relationship that has shaped the course of human history, a relationship that confirms and complements our comprehension of the evolution of civilization. Through this model, we gain a deeper appreciation of the intricate dance between technological advancement and the ever-evolving tapestry of human values.

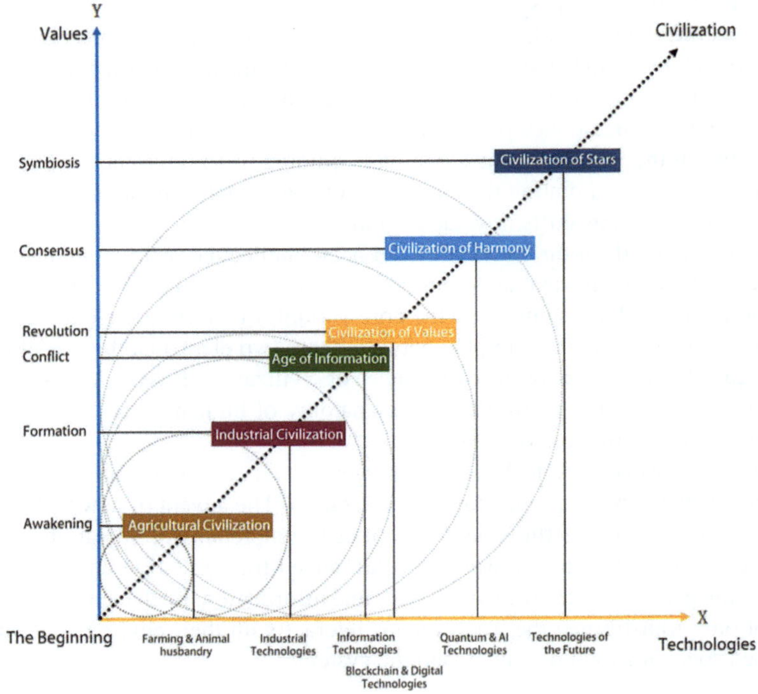

4.1.1 Chart Structure

The chart builds a two-dimensional coordinate system that clearly presents the interactive architecture of technology and value.

Coordinate Axis: The horizontal axis focuses on the "value" dimension, finely dividing from left to right into "yuan point", "value awakening", "value shaping", "value hedging", "value revolution", "value consensus", "value symbiosis" and "civilization", just like a clue to the growth process of human value. The vertical axis anchors the "technology" field, and marks "farming and grazing technology", "industrial technology", "information technology", "block chain", and "digital technology", "quantum and artificial intelligence technology", "future technology" in order, like a chronicle of the progress of human technology.

Curves and Points: Multiple concentric elliptical curves in the chart emerge elegantly, they are just like the ripples in the long history, each representing the different stages of civilization development, layer by layer nesting, showing the gradual evolution of civilization from low to advanced. Each elliptical curve is accurate. These points are like stars, which are closely connected with the corresponding value stage on the horizontal axis through clever dotted lines, weaving a link between the interaction between technology and value. More subtly, Points of different colors are given a unique type of civilization: brown dots stand still, A symbol of "agricultural civilization", Located exactly in the intersection coordinate of farming technology and value awakening, Tell the initial value of human beings from water and grass to pastoral settlement; The purple dot shines on, For the term "industrial civilization", Located prominently at the crossroads of industrial technology and value shaping, Witnessed the surging process of the values being reforged under the industrial tide; The green dot is vibrant, For the "information age", Vitality blooms in the coordinate interval of information technology and value hedging, Mapping the noise of the collision of multiple values; The orange dot shines brightly, Identify the "Value-Based Civilization ", At the intersection of the block chain and the digital technology and the value revolution, To cover the subversive reshaping of traditional value system by digital frontier technology; bluePoint deep and lofty, meaning "great civilization", resting in the key point of quantum and artificial intelligence technology and value consensus, looking forward to the future high-tech context of human understanding value harmonious vision; dark blue point mysterious, pointing to "interstellar civilization", hidden in the vast distance of future technology and value symbiosis, outline the ultimate dream of human beings towards the universe and collaborative evolution with everything.

4.1.2 Graphic Meaning

4.1.2.1 The Relationship Between Technology and Value

This chart vividly shows a magnificent epic of technological progress and value change walking hand in hand. From the germination of the most simple farming and animal husbandry technology to the future technology vision full of infinite imagination, every technological leap-off accurately corresponds to the profound changes in the value field. Technology, like a pair of invisible but powerful hands, continues to promote the awakening of value from ignorant, after carving, fierce hedging, old revolution, towards broad consensus, and finally reach the harmonious coexistence, and then pull human civilization up the ladder and climb to a higher peak.

4.1.2.2 The Evolution of Civilization

Agricultural Civilization: Under the warmth and nourishment of farming and grazing technology, human beings gradually get rid of the nomadic wandering and open a new chapter of settled farming. At this time, values such as being dependent on the land, family blood, reverence and respect for natural gods were quietly awakened, becoming the spiritual foundation for maintaining social order and ensuring the reproduction of ethnic groups, marking the first step of primitive accumulation of human civilization.

Industrial Civilization: The advent of industrial technology, like a loud bang of The Times, has completely rewritten the trajectory of human development. Factory chimney, machines, machine, the wave of urbanization, in the process of industrialization, the pursuit of efficiency first, encourage innovation breakthrough, advocating individual struggle value orientation arises at the historic moment, the value system of human society is deep remodeling, toward the direction of more expansionary and competitive.

Information Age: the emergence of information technology, like an instant invisible web in the world, information like lightning through it. However, under the torrent of information, the values of different regions, cultures and classes collide and hedge each other. On the one hand, the voice of free sharing of information and global interconnection and cooperation is growing, becoming the strong voice of The Times; on the other hand, the emerging demands of personal privacy protection and digital ethics are also like undercurrent, which need to be responded to, and human civilization has entered a complex transition period full of contradictions and opportunities.

Value-Based Civilization: When block chain and digital technology are emerging, a shocking value revolution has quietly begun. Digital technology with its decentralization, traceability, highly encryption features, based on the traditional authority, centralized control value system, the virtual world booming, new value judgment standards, in the digital scenarios such as yuan universe, social, economic, creative logic is written, human civilization opens the digital value new exploration of reconstruction.

Datong Civilization: Looking forward to the future of quantum and artificial intelligence technology, human beings are expected to use these super technological forces to break through many practical difficulties and achieve unprecedented global collaboration. In a highly developed science and technology ecology, people of different races, countries and beliefs will gradually gather and form a more unified and widely recognized value consensus based on the global challenges and opportunities they face, steadily stride toward the ideal realm of "the world", and inject soul strength into the sublimation of human civilization.

Interstellar Civilization: Standing in the present and looking into the distance, the mystery of the future technology has not yet been fully unveiled, but it is enough to attract the imagination. When human master interstellar navigation, alien resources development, cross-galaxy communication science and technology, the earth will no longer be the cradle of loneliness, human will blend with the universe depth, value, build up a cover interstellar dimension of grand civilization community, epic across from the earth civilization to interstellar civilization, this is the ultimate vision of the development of human civilization.

4.1.3 Conclusion

In a word, this chart is like an accurate key. It serves as a detailed guide that unlocks the mysteries of human progress, mapping out the intricate relationship between technological advancements and the evolution of human values. Through the delicate outline of this dynamic correlation, the chart traces the simple origins of human civilization, starting from the ancient agricultural societies where the plow and the seed were the harbingers of change. It charts the course through the industrial revolution, where steam engines and assembly lines transformed societies and economies, to the digital age, where computers and the internet have woven a global web of interconnectedness. The chart extends its vision even further, all the way to the bright future of the vast interstellar civilization, where humanity may 1 day transcend the confines of our home planet and explore the cosmos.

With intuitive and shocking visual language, the chart captivates the viewer, presenting a vivid narrative of technological progress as the source power that has propelled humanity forward. It illustrates how technology has continuously shaped and rewritten human values in the long history of our species. For instance, the invention of the printing press democratized knowledge, challenging the authority of the church and the ruling elite, and paving the way for the Renaissance and the Enlightenment. Similarly, the chart shows how the advent of electricity and the internal combustion engine not only changed daily life but also reshaped societal norms and economic structures.

The chart emphasizes that technology is not just a tool; it is a force that drives the vicissitudes of civilization and its continuous upgrading. It clearly reflects the tortuous development trajectory of human society under the impact of different technological waves. Each wave, from the steam engine to the internet, has brought about profound changes in how we live, work, and interact. The chart highlights the industrial revolution's transformation of the workforce from agrarian labor to factory jobs, and the information revolution's shift towards a service and knowledge-based economy.

Moreover, the chart delves into the ethical and philosophical implications of technological progress. It questions whether the relentless pursuit of innovation has come at the cost of our environment, our privacy, and our social fabric. It provokes thought on how we can harness technology responsibly to ensure that it serves the greater good and does not exacerbate inequality or lead to unintended consequences.

In conclusion, this chart is not merely an illustration; it is a comprehensive narrative that encapsulates the essence of human ingenuity and ambition. It serves as a reminder that as we stand on the cusp of new technological frontiers, we must learn from the past to navigate the future. The chart is a call to action for policymakers, technologists, and citizens alike to collaboratively steer the course of human civilization towards a sustainable and equitable future.

4.2 Model Building and Association

4.2.1 Technology: The Driving Force of Human Civilization

Technology, as the core engine of human civilization, continues to output surging power, driving the progress of our society forward. Looking back at the starting point of history, the advent of farming and grazing technology marked the dawn of a new era for mankind. This technological breakthrough led humanity to bid farewell to the wild and nomadic lifestyle, take root in the land, and open a new chapter of agricultural settlement. With the domestication of livestock and the sowing of grain, a fundamental change in the mode of survival was realized. People began to cultivate the land, forming stable communities and laying the foundation for the development of civilization.

As centuries passed, industrial technology emerged in succession, igniting the raging fire of the industrial revolution. The invention of the steam engine and the widespread use of electricity marked significant milestones. Large-scale factory production swept across the globe like a prairie fire, transforming the urbanization process and reshaping the social structure and economic patterns. The development rate of human civilization experienced a sharp increase, as the industrial age brought about unprecedented advancements in productivity and innovation.

Entering the information age, the Internet emerged as a magical sky net, breaking the shackles of time and space in information dissemination. Knowledge could now be displayed and shared on a global scale, revolutionizing the way we work, live, and interact socially. The digital revolution has ushered humanity into the fast track of digital survival, where information flows freely and connectivity knows no bounds.

Today, cutting-edge technology clusters such as blockchain, digital technology, quantum computing, and artificial intelligence are reshaping the world with an overwhelming momentum. Taking digital technology as an example, it is like a global storm, invincible and deeply penetrating the social texture. It is changing the economic model, with the virtual economy and virtual land transactions becoming increasingly popular. Digital asset sales in the metaverse are booming, creating new economic opportunities and transforming traditional financial systems.

Furthermore, digital technology is reshaping the social paradigm, making remote immersive social interactions a daily occurrence. The Virtual Party in Decentraland is a testament to this, where people from all over the world come together in a digital space to celebrate and interact, transcending physical boundaries. This new form of socialization is redefining community and connectivity.

In the realm of work, digital technology has created a new mode of remote collaboration, significantly reducing enterprise costs while boosting efficiency. Platforms like Microsoft Mesh for Teams are making it possible for global teams to work seamlessly together, regardless of geographical location. This shift has not only transformed the workplace but also opened up new possibilities for work-life balance and global talent acquisition.

In conclusion, technology continues to be the driving force behind human progress, constantly evolving and reshaping our world in profound ways. From the dawn of agriculture to the industrial revolution, and now into the digital age, each technological leap has propelled humanity into a new era of innovation and possibility.

4.2.2 Value: The Core of Human Civilization

Value, which runs through the long development of human civilization, is its rock-solid core. Throughout the ages, the concept of value has been the driving force behind the progress and evolution of societies. During the Agricultural civilization period, farming technology was the backbone of society. The value of land, family loyalty, and the fear of natural gods were plain and simple. People worked hard on the land, families banded together, and every farming community offered pious prayers for a bountiful harvest. These values tied people firmly to their land and maintained social stability, creating a strong centripetal force within groups.

With the surge of industrial civilization, values such as the pursuit of efficiency, innovation, and individual struggle became the beacon of The Times. These values perfectly fit with the needs of industrial mass production, encouraging countless pioneers to participate in technological innovation and market expansion. This led to rapid economic growth and the creation of industrial brilliance. Factories sprung up, cities expanded, and the world was transformed by the steam engine and electricity, all underpinned by these new values.

As we entered the information age, the flood of information collided with multiple values. On the one hand, the free sharing of information and global interconnection and cooperation became the universal pursuit, breaking down barriers of regional knowledge and cultural isolation. The internet emerged as a powerful tool for communication, education, and the dissemination of ideas. On the other hand, personal privacy protection, digital ethics, and other emerging demands surged to the forefront, presenting new challenges that humanity had to address. The digital age became a turbulent value revolution, where the emergence of the value twin—the fusion of technology and human values—began to shape new value judgment standards in the virtual world.

In the field of ecological protection, the value of sustainable development was upheld, and digital technology was used to accurately monitor the environment, track climate change, and manage natural resources more efficiently. Satellites and sensors provided real-time data, enabling scientists and policymakers to make informed decisions to protect our planet.

In the public health battlefield, the value of the community of human destiny became evident. Countries began to share anti-epidemic information and engage in transnational rescue efforts. The COVID-19 pandemic highlighted the importance of global cooperation, as vaccines were developed at unprecedented speeds and distributed across borders. Cultural exchange stages became virtual, connecting people from different parts of the world in ways that were previously unimaginable.

The digital age has also seen the rise of social media platforms, where individual expression and collective action have created new value systems. Influencers and activists use these platforms to raise awareness about social issues, driving movements that challenge traditional norms and advocate for change.

In the commercial sector, the value of customer-centricity and data-driven decision-making has transformed the way businesses operate. Companies now leverage big data to understand consumer behavior, personalize services, and create innovative products that cater to the needs of a digital-first audience.

In the realm of creativity, the value of intellectual property and originality has been both challenged and celebrated in the digital age. Artists and creators have new avenues to showcase their work, but they also face issues of piracy and copyright infringement. The balance between accessibility and protection of creative works continues to be a topic of debate.

As humanity stands at the crossroads of value choice, it is clear that the digital age is not just a technological revolution but a profound transformation of our values and the way we live. The challenge for society is to navigate this turbulent value revolution, ensuring that the new value systems that emerge promote the well-being of individuals and the collective good of humanity.

4.3 Human Civilization Function Formula (HCF Formula)

4.3.1 The Complete Expression of the Formula

Human civilization is like a complex and exquisite function system. We may set technology (T), value (V), negative entropy (η) as core independent variables, while civilization achievement (C) is taken as the closely corresponding dependent variable, thus constructing a dynamically related function relationship:

$$C = \sum\nolimits_{\mathrm{i}=1}^{5} \left(T_i \cdot \frac{V_i}{\max(V_i)}\right) \cdot \eta$$

4.3.2 Symbols and Numerical Definitions

Symbol/Num	Name	Definition and explanation
C	Civilization index	A comprehensive quantitative metric measuring the progress of human civilization, ranging [0, 100] Example: C = 75 indicates a moderately ordered civilization state
$\sum_{i=1}^{5}$	Five-dimension	Weighted summation across five dimensions: Peace, development, equity, ecology, and cultural diversity Derived from Yan Lijin's theory, a dimensional reduction abstraction of the UN SDGs
T_i	Technical variable	Reflects the contribution level of technological innovation to the i-th dimension, range [0, 1] Calculation: $\mathrm{T_i} = \frac{\textit{Number of Domain Patents} \times \textit{Technology Adoption Rate}}{\textit{Global Patent Total} \times \textit{Average Adoption Rate}}$
V_i	Value variable	Raw quantitative value of the i-th dimension (e.g., peace index = 75), range [0, 100] Data sources: UCDP conflict data, NASA satellite remote sensing, etc.
$\max(\mathrm{V_i})$	Historical maximum	Historical peak value of the i-th dimension (e.g., historical maximum peace index = 85) Purpose: Normalizes $\mathrm{V_i}$ to eliminate unit differences
$\frac{V_i}{\max(V_i)}$	Value variable	Maps raw values to [0, 1] for standardization Example: $\frac{75}{85} \approx 0.8828$

(continued)

(continued)

Symbol/ Num	Name	Definition and explanation
η	Technology-value	Reflects synergy between technology and values, range [0, 1] Calculation: $\eta = \frac{\sum_{i=1}^{5}\left\|T_i - \frac{V_i}{\max(V_i)}\right\|}{5}$

4.3.3 Dimension Weights and Variable Mapping

Dimension	Weight (%)	Technical variable (T_i)	Value variable (V_i)
Peace	30	Application rate of quantum encryption in conflict mediation (e.g., quantum communication for secure negotiations)	Ratio of social media peace sentiment (via twitter API analysis) Global conflict events (UCDP database)
Development	25	Blockchain patents in cross-border payments × adoption rate	GDP per capita (World Bank) + education investment (UNESCO) + vaccination rate (UNICEF)
Equity	20	Coverage of digital ID in financial inclusion (e.g., unbanked population using digital IDs)	Global gender gap index (WEF) + tax transparency (tax justice network)
Ecology	15	Carbon capture patents × adoption rate (e.g., carbon sequestration projects)	Forest cover (NASA satellites) + CO_2 emissions (IEA) + biodiversity index (WWF)
Cultural	10	User penetration of metaverse in cultural preservation (e.g., virtual museum visits)	UNESCO heritage funding + language diversity index (Ethnologue)

4.3.4 Function Relation Parsing

Technological Progress (Δ T) is like the magic "master of slope-adjustment" of the function curve. Every milestone technological breakthrough is like a stroke on the historical canvas, subtly changing the slope of the development trajectory of civilization and accelerating civilization like the arrow away. For example, the emergence of industrial technology, like a dazzling dawn breaking through the night sky, instantly pulls the human society out of the leisurely rhythm of the agricultural era, leaping into the high-speed track of the industrial era, opening a new chapter of development.

Value (V), as the soul core value of civilization, is just like a forward-looking navigator, accurately anchoring the development direction of civilization. Different value orientations (Δ V) are like a baton, cleverly guiding the direction of technology development and application. If we firmly uphold the value of sustainable development in the field of ecological protection, technology research and development will focus on clean energy, environmental protection materials and other frontier areas, for the earth home escort; on the contrary, if short-sighted, excessive pursuit of short-term economic interests, may lead to a series of thorny problems such as excessive development of resources, ecological environment deterioration, as if buried obstacles for the road of civilization, hinder the pace of development.

Negative entropy (η) is like an indestructible fortress, ensuring the stability of the functional system against all kinds of interference from inside and outside. The continuous input of negative entropy factors, such as knowledge inheritance, system improvement and scientific and technological innovation, is like injecting a steady stream of "stable energy" into the civilized system, maintaining the civilized system in a low entropy and orderly state, and ensuring that technology and value can work together to play a harmonious movement of civilization development.

When the three form a benign interaction, technology enabling value concept in the real world, just like the pursuit of fair technology to promote the benefit of the people; the negative entropy solid system foundation is not shaken, human civilization is like a thriving tree, climbing toward the sky, showing infinite vitality and vitality. When technology is harnessed to support and enhance the values that society holds dear, it can lead to remarkable advancements. For instance, consider the way medical technology has evolved to prioritize patient well-being, resulting in longer, healthier lives for millions. The concept of negative entropy, or the idea of maintaining order and complexity in systems, ensures that the foundational structures of society remain robust and unshaken. This is akin to the way ecological systems maintain balance, allowing human civilization to flourish like a tree that reaches for the sky, its branches extending and growing, symbolizing the endless potential and vigor of human progress.

On the contrary, if the three imbalance, civilization may fall into the predicament of stagnation or even decline, facing many crises. The delicate balance between technology, values, and system stability is crucial. If this balance is disrupted, the consequences can be dire. For example, consider a society that advances technologically without regard for ethical considerations, potentially leading to misuse of technology and harm to society. Similarly, if the systems that underpin civilization are not maintained and allowed to decay, the entire structure of society can become unstable, resulting in stagnation or decline. This imbalance can manifest in various crises, such as economic downturns, social unrest, or environmental disasters, all of which can threaten the very fabric of civilization.

To sum up, technology, value and negative entropy are closely linked and complementary, and together constitute the core mechanism of the dynamic evolution of human civilization. The synergy between these three elements is what propels human civilization forward. Technology must be guided by ethical values and work in harmony with the principles of maintaining system stability. Only with a deep insight, accurate understanding and a firm grasp of the relationship between them can we clear the fog of history, have a clear insight into the development of human civilization, and lead mankind to a more brilliant future. Understanding this intricate relationship allows us to navigate the complexities of the past and present, providing a clear vision for the future. It empowers us to make informed decisions that will foster a sustainable and prosperous civilization, guiding humanity towards a future that is not only brighter but also more equitable and resilient.

Part II
"Questions of the World" and the Value Revolution of Human Civilization Brought About by the Digital Era

Chapter 5
"What's Wrong with the World"?

In today's time, the world is like a turbulent, turbulent, treacherous and dangerous adventure. It is a vast and complex tapestry woven with the threads of countless cultures, economies, and political systems, each contributing to the dynamic and often unpredictable nature of global events. In the process of surging, many difficult problems emerge as layers, like sedimentary rocks revealing the history of the earth. After the surging waves of change and conflict, with an overwhelming, overwhelming majestic momentum, heavy shrouded in the challenge of human civilization on channel, like a tight, suffocating dark net, let forward ships confused, lost, and like a shelter, heavy gloomy fog, make people in exploring the future to deep confusion, lead into deep confusion: what's wrong with the world?

When we are full of the human destiny community that hot and deep concern, we find ourselves grappling with questions that seem to have no easy answers. We are deeply exploring the root behind the complex situation, peace deficit and security deficit, development deficit, governance deficit, civilization deficit and system hegemony and a series of serious and urgent problems. These issues are like the towering male peaks in the human journey to a brighter future, towering, steep, each like an insurmountable barrier, mercilessly blocking the firm pursuit of progress and happiness. They make it difficult for mankind to walk on the bright road of development.

For instance, the peace deficit is evident in the ongoing conflicts and wars that plague various regions, leaving behind a trail of destruction and human suffering. The security deficit is palpable in the rise of cyber threats and terrorism, which cast a shadow of fear over the global community. Development deficits are seen in the stark contrast between the prosperity of some nations and the poverty of others, a gap that continues to widen. Governance deficits are apparent in the failure of some governments to effectively serve their people, leading to corruption and instability. The civilization deficit manifests in the erosion of cultural heritage and values, as globalization often imposes a homogenizing effect on diverse societies. Lastly, the

L. Yan, *Value-Based Civilization*, Understanding China,
https://doi.org/10.1007/978-981-96-9490-7_5

system hegemony perpetuates an imbalance of power, where certain nations dominate international affairs, often to the detriment of smaller or less powerful states.

These problems, intertwined and complex, require a concerted effort from the global community to address. It is only through cooperation, dialogue, and a shared commitment to the principles of justice and equity that we can hope to navigate through these treacherous waters. As we strive to unravel these knots of contention, we must also foster a spirit of understanding and empathy, recognizing that the challenges we face are not just the problems of individual nations, but of humanity as a whole. Gradually, from behind the appearance of chaotic confusion, we can begin to discern a path forward, one that leads not just to a brighter future, but to a future where all of humanity can thrive together.

5.1 Peace Deficit and the Security Deficit

Peace, without doubt, is the most solid foundation for the stable survival and prosperity of human society, just like the solid foundation of a magnificent building buried deep in the ground. Once this foundation is shaken, the whole building will be in danger and instantly collapse into ruins. However, the reality of the scene is like a sharp, cold light sharp blade, mercilessly stabbing the heart of every person with conscience. Looking at today's world, many places filled with smoke and war, the thorny problem of peace deficit is more striking and dazzling highlighted, like a "disaster" in the night sky, indicating the "disaster star", has brought endless suffering and pain to human beings. Local conflicts like a long cure, difficult to pull the ills chronic, in parts of continuous crazy, spread, for a long time to stop, as the wilderness on the tenacious weeds, even in the burning, but still burn, blowing and raw, stubbornly rooted in the piece of suffering land. The Middle East is undoubtedly a typical representative of the mire of war. The people of Syria, Iraq and other countries have long suffered the heavy suffering brought by the war. Once warm and beautiful home in the relentless bombardment of gunfire into a broken wall, that broken. The bricks and stones seem to silently tell the peace and harmony of the past and now desolate desolate; relatives displaced, life and death, the streets and lanes often echoed the lost relatives that tore heart crack lung, heartbroken cry, like the soul of the bitter cry, lingering for a long time. Innocent life in bullets vulnerable like ants, fleeting, their dreams, vision of hope and the future, by the war rolling smoke, the most basic survival rights and interests are difficult to protect, even a clean drinking water, a can shelter tile become a distant hope, people can only struggle to survive in the ruins, days like years, every minute every second is Fried.

Many factors, such as ethnic contradictions, religious disputes, complex and fierce games in geopolitics, are like a group of chaos and endless chaos, complicated and closely intertwined, making the parties in the conflict deeply locked and difficult to extricate themselves, and the dawn of reconciliation is like looking for a needle in the sea. Due to historical grievances, cultural differences, territorial disputes and other issues, the conflict has a long history, as bitter and incompatible, religious beliefs like a deep and difficult gap, in the flames of conflict, and the conflict for their own narrow interests. Forces in the game of chess stalled, the hope of peace is repeatedly by the war thick haze relentless shadow, regional situation continued unrest, brought the local people, also for the global peace prospects covered with a layer of thick gray, despair, like a huge, airtight black cloth, will hope light shade, let the whole world into a abyss of darkness and despair.

Together with it, the security deficit is just too severe to be ignored. Before the cloud of traditional security threats has been completely dissipated, the new security challenges have come like storms. Terrorism is like a ghost hidden in the dark, wandering in every corner of the world, with extreme violence, inhuman cruel means deliberately create panic atmosphere, in an attempt to disturb the social order. Whether busy city traffic, neon flashing busy streets, or personnel bustling, laughter of public gathering places, at any time is likely to become terrorist attacks, unarmed innocent civilians become the tragic targets of wanton attacks, they were immersed in the joy of life warmth, the next second laughter may instantly be deafening explosion and sad grief cry instead, fresh life stopped, leaving only relatives grief, heartbroken cry. Once rampant "Islamic State", in the Middle East, chilling and chilling, bloody slaughter, merciless looting, city streets, the bodies of innocent people scattered everywhere, to regional and even global security. The situation has caused an extremely violent shock, like a strong earthquake, shattered peoples vision of peace and security, so that the fear of fear in everyone's heart.

At the same time, with the rapid development of information technology at a rapid speed, cyberspace has evolved into a new battlefield without smoke but crisis, and the network security problem is increasingly severe and complex, difficult and difficult to do. Hacking, information leakage, network fraud malignant incidents frequently, not only for personal privacy and property security pose a direct threat, like a hanging overhead, may fall at any time, is the national key infrastructure, financial system, national defense security crucial areas, formed a significant and potential risks, like a hidden in the shadows, may detonate the bomb at any time. Multinational crime is increasingly rampant, unscrupulous, smuggling, drug trafficking and trafficking illegal acts wanton across borders, woven into a complex, intertwined criminal network, seriously disrupted the normal social order, ruthlessly erode the security foundation of human society, like a group of greedy termites, slowly empty social stability "wall", let the whole society crumbling, precarious, may at any time into chaos and collapse.

There are inextricably linked and closely intertwined internal links between the peace deficit and the security deficit. The two interact and influence each other, like a heavy chain of a vicious circle, constantly biting the stable foundation and peaceful environment of human society. The outbreak of war was like a storm, destroying urban infrastructure, bumpy roads, water and electricity supplies, communications networks, jobs, struggling with hunger and poverty, education and health systems, children lost the precious opportunity to learn knowledge and pursue dreams, patients without timely and effective treatment. These chaotic situations created a huge amount of unrest, creating a hotbed for terrorism and other security threats; increasing the tension in the region, overwhelming the high wall, forcing humanity to struggle in the pursuit of peace and security, and cost countless families Broken, relatives are separated, social development stagnation, the economy into a deep predicament, as if into a long and endless winter, cannot see a glimmer of hope.

5.2 Development Deficit

Development, from beginning to end, is the core theme of human society, and is the key master key to open the door to solve many problems. It is like a golden key to open the door to a prosperous and better future, and lead mankind to a bright road. However, it is regrettable that the current global development imbalance is extremely prominent, like a huge and bleak planet in the night sky, striking in front of the world, let people be daunting, like a towering, insurmountable steep mountain, brutally blocking the pace of human progress. Focus on economy, the north and south gap is still like a deep gap, developed countries rely on first mover advantage, firmly control the dominant staff of the global economy, to enjoy the high profits and high quality resources rich dividends, as standing on the top of the world overlooking the giant beings, enjoy scenery glory, the worlds wealth and resources almost greedy. In stark contrast, the vast number of developing countries are trapped in many difficulties, like climbing in the dark valley bottom, every step is full of hardships and challenges.

Infrastructure construction is weak, transportation, energy, communications and other key areas of the lag situation, like an invisible but heavy shackles, tightly restricting the pace of economic take-off. In some remote areas, the roads are rough, muddy and bumpy, it is like a boat sailing in the rough, undercurrent sea, bumps, downs and dangers, transportation of goods is difficult, often delayed because of poor roads, lead to a significant increase in costs; unstable energy supply, frequent power cut and water cut, enterprise production was forced to interrupt and stall, life are also in chaos, the night can only be spent in the darkness; communication signals are weak or even completely absent, people look like isolation, information exchange is severely blocked, companies are difficult to settle down, economic development is out of the question, like an island forgotten by the world, floating alone outside the tide of The Times. Shortage of funds makes the plight of the industrial upgrading difficult, if in the mire, difficult to break free from the shackles of low-end manufacturing, in the global industrial chain in low added value, low profit low position, can only provide cheap raw materials and labor, itself is difficult to realize the transformation and upgrading of industry, obtain higher economic benefits, if was tightly bound by an invisible rope, unable to break free. Many countries on the African continent is a typical example, although the land is extremely rich in natural resources, like a undeveloped treasure library, but because of the lack of necessary development technology and sufficient funding, struggling in the poverty, people living poor, basic public services such as education, medical care, and the rich scene of developed countries form a sharp contrast, if two from different time and space, different world, is a prosperous advanced city, one is broken backward countryside, gap painful, sigh.

Environmental issues are also like a deafening and heavy wake-up call, bringing great challenges to global development, and have become an indispensable part of the development deficit. With the process of industrialization and urbanization accelerating like a high-speed train, human excessive exploitation and unreasonable use of natural resources, like a runaway and crazy plunder, causing a series of shocking and painful ecological crises. The forest was heavily cut down, the rapid reduction of the green vegetation, once lush, vibrant forest now remains but bare stumps, like a lonely watch guard, silently witnessing the vicissitudes of the ecology; soil erosion phenomenon is becoming increasingly serious, the fertile land is gradually desertification, the howling wind went by, roll up the dust, as if the end of the world is coming; a sharp decline in biodiversity, many rare species are endangered, animals have lost their habitat, ecological chain fracture and fragmentation, the ecosystem of Earths home is suffering from unprecedented destruction, like a once beautiful garden was brutally and brutally trampled, to become desolate, desolate, and unrecognizable. Global warming is like a rickety sword of Damocles, with accelerated melting of glaciers, continuously rising sea levels and extreme climate events frequent outbreaks, not only mercilessly threaten the residents of coastal areas home, their houses may be submerged by water, the pastoral water, people forced to be displaced and displaced; also like a heavy blow, to agriculture, fishery and other traditional industries, food production, fishery resources exhaustion, farmers and fishermen work hard for a year without grain, seriously endanger food security and economic stability, let human survival into the crisis, struggling. Developing countries due to their own development needs is imminent, often in response to environmental problems into an awkward situation, to helpless early industrialization legacy from the developed countries, and in their own development and environmental protection dilemma of difficult balance, struggling to find way out, if in thorns, full of spike jungle looking for the faint hope, every step with blood and pain, hard.

In addition, there are also many worrying and anxious shortcomings in the field of social development. Education inequality like a dazzling striking, difficult to heal cracks, widely and stubbornly exist, high quality education resources like rare treasure, highly concentrated in a few developed countries and regions, and a large number of children because of poverty of developing countries, deprived of the right to get a good education opportunity, which further imprisoned their future development potential, like a invisible but fatal curse, formed the intergenerational transmission of poverty. In some poor areas, the classroom dilapidated, mottled walls, tables and chairs incomplete, lack of teachers are weak, teachers only with vigor, teaching knowledge under the condition of arduous, the children can only in the dim flickering light, with crude rough school supplies, hard to learn knowledge, and children of developed countries enjoy the modern teaching facilities and rich variety of education resources, spacious and bright classroom, advanced intelligent multimedia equipment, professional teachers team, contrast, sigh, emotion. The problem of uneven distribution of medical resources is also severe and urgent. Many poor areas are rampant, and the plight of the lack of medicine is like an insurmountable barrier for the people. There is a huge gap between the average life expectancy and the developed countries. People often lose their lives because of a minor illness, and life is so fragile in the face of the disease. Employment difficulties is like a global haze fog, shrouded over the countries, especially under the heavy economic downward pressure, high youth unemployment, a large number of labor like idle waste resources are wasted, they with dreams and enthusiasm, but cannot find talent stage, not only seriously affect personal livelihood and family happiness, also like a bomb may detonate at any time, bring great hidden danger to social stability, social increasingly acute contradictions, conflict is hair-trigger. These development in the process of unbalanced, uncoordinated, unsustainable problem superposition of each other, like a crumbling, dangerous building, makes the global development deficit increasingly heavy, seriously hindered the human society overall forward firm pace, like a carriage stuck in the mud, no matter how to struggle, are difficult to break free, move on, can only wander in situ, anxiously waiting to save.

5.3 Governance Deficit

In the tide of globalization, if surging tide, rolling forward, unstoppable present, the links between the world increasingly closely, like a closely intertwined, intricate network, interdependence climbed to new heights, global problems like the tide recede after bare ferocious rocks, emerge in endlessly, all urgently calls for a set of efficient, collaborative global governance system debut, if sailing in the dark of the sea ships need a bright beacon direction, for lost ships light up the way forward. However, the reality is disappointing. While the current global governance is deeply in the predicament of serious governance deficit, like a lost ship, drifting in the vast sea, drifting with the flow, I do not know where to go, like a lost lamb, helpless.

On the one hand, the existing international organizations and governance mechanisms are like an old, rusty and ineffective machine, and there are many obvious and criticized defects. Although the international system built with the United Nations at its core has played an important role in maintaining world peace and promoting international cooperation, it is still struggling and struggling to face some new problems and thorny challenges. Decision-making mechanism of permanent security council, in some complicated, about the fate of global international issues, if in the wheel of the mire, difficult to reach a consensus of efficient, unified, based on the fierce game between powers, often lead to key decision difficult, delayed the golden time to the crisis, as at the scene of the fire, all the debate who action first and missed the best time, the fire spread out of control, cause greater losses. The world trade organization (WTO) shoulder the sacred mission of promoting global trade liberalization, but in recent years its multilateral trade negotiations like deadlock, trade protectionism like a upstream, strong rise, countries have built towering trade barriers, like a strong line, seriously hindered the health and smooth development of global trade, Has greatly weakened the authority of WTO in the formulation and implementation of international trade rules. If an authoritative judge is questioned for its fairness and is unable to enforce the law normally, leading to frequent trade disputes and chaotic global trade order. The international monetary fund (IMF) and the world bank in the field of international financial governance, but in the capital allocation, loan conditions, like pairs of unfair hand, there are many unreasonable, failed to fully take into account the actual needs and difficulties of developing countries, causing controversy and discontent, thus affect the play as a global financial stabilizer function, like a reliable shield appeared loopholes, unable to provide a solid guarantee for the global financial system.

On the other hand, the participation of global governance subjects is like an unbalanced balance and seriously unbalanced. With their strong economic and political strength, developed countries are the leading role in the center of the stage and dominate the process of global governance. They often put their narrow interests first and turn a blind eye to the rights and interests of developing countries, even at their expense. Developing countries although numerous, like a vast expanse of stars, and global problems, but in the international rule making, decision process discourse like weak dim candlelight, serious shortage, difficult to fully, smoothly express their demands, can only be forced to become a passive recipient rules, which further exacerbated the global governance system of unfair and unreasonable, like a game, one party has absolute jurisdiction, the other party can only silently endure unfair decisions, suffering, full of injustice. In response to climate change the global urgent issues, the early industrialization of carbon emissions like a heavy history, is one of the main culprit of global warming, but on the stage of the international climate negotiations, some developed countries like depend on children, reluctant to bear the corresponding historical responsibility, trying to shuffle wrangling, in the target setting. In terms of financial and technical assistance, it is extremely stingy, making it difficult for developing countries to tackle climate change and act effectively, thus affecting the global climate governance process.

The consequences of this governance deficit are comprehensive. In public health, the global pandemic of COVID-19 is a heavy wake-up call. When the outbreak came suddenly, countries around the world should work together to overcome the difficulties, but the reality is that the international coordination mechanism is slow and chaotic, and the unfair distribution of vaccines has become a prominent problem. Developed countries with abundant financial and technical strength, a large number of hoarding vaccine, like selfish miser hoarding treasure, and many developing countries can only hope "miao", waiting, lead to outbreak in these areas, life, economic further hit, the global resistance to disease road become bumpy, struggling. In terms of network governance, due to the lack of unified and strong norms, network crime, information leakage, cross-border tracking and blow difficulty, to the national security, economic development and personal rights and interests of people have brought a serious threat, if the network world into the lawless "wild west", chaos.

5.4 The Civilization Deficit

Civilization, as a unique and bright spiritual symbol of human society, should light up the night sky like the stars, bringing a gorgeous picture of harmonious coexistence and shared beauty to the world. It is the beacon that guides humanity through the darkest times, a testament to our collective achievements and aspirations. Civilization is the mosaic of human creativity and wisdom, a tapestry woven from the threads of diverse cultures, languages, and traditions. It is the repository of our greatest art, literature, philosophy, and scientific discoveries. It should be a source of inspiration and pride, a shared heritage that unites us in our common humanity. However, the current reality is very worrying, the civilization deficit is like a thick haze, quietly shrouded in the world. It obscures the vision of a peaceful and prosperous future, where misunderstandings and prejudices take root, and even conflicts between different civilizations occur from time to time, like thorns on the road of communication, hindering the approach of the human mind.

The world today is fraught with challenges to the very essence of civilization. Some western forces have long upheld the principle of "civilization superiority". They look at other civilizations through the lens of their own cultural biases, as if they were arrogant aristocrats, trying to promote their own values and development models to the world, lacking basic respect and tolerance for heterogeneous civilizations. This hegemonic attitude has caused many regional aversions and conflicts. In the Middle East and Central Asian countries, the west has forced the so-called "democracy", regardless of the local history and culture, social structure and public will. This has often led to social unrest, wars, and people living in hot water. Once ancient and splendid civilizations have been covered by war smoke, cultural heritage has been destroyed, and precious historical memory has been ruthlessly erased. This is a huge loss for human civilization, a loss that cannot be measured in mere physical destruction but also includes the erasure of cultural identity and historical continuity.

The legacy of these conflicts is not just the physical devastation of cities and the loss of innocent lives, but also the psychological scars that are etched into the collective memory of societies. The damage to cultural heritage is not just the loss of buildings and artifacts, but the severing of the connection between people and their past. The destruction of historical sites is not just a loss of tourism revenue, but a loss of the educational opportunities that these sites provide. The erasure of languages and dialects is not just a loss of communication tools, but a loss of the unique ways of thinking and expressing that they represent.

At the same time, the phenomenon of unequal cultural exchange, where some cultures are promoted and others are marginalized, continues to be a pressing issue. This imbalance is evident in the global media landscape, where certain narratives dominate and others are relegated to the sidelines. It is seen in the international book market, where translations from dominant languages flood the shelves, while works from smaller languages struggle to find a voice. It is felt in the arts, where certain styles and genres are celebrated on the world stage, while others remain hidden in local communities. This inequality in cultural exchange not only limits the diversity of human expression but also hinders the full understanding and appreciation of the richness of our global heritage.

5.5 Institutional Hegemony

On the international stage, institutional hegemony is like a hidden but destructive "behind the scenes", controlling the development trend of many key fields and casting a gloomy shadow on fairness and justice in the world. This invisible force operates in the shadows, manipulating the strings of global politics and economics. A few developed countries, led by the United States, with their strong comprehensive strength, force their own clauses and rules in the process of building international economic, political and military systems, like an unfair game. They are both players and judges, and control the balance of the game. Their influence extends far and wide, from the corridors of power in Washington to the boardrooms of multinational corporations, shaping policies and decisions that affect billions of lives across the globe.

In the economic world, the dollar's hegemony is a classic example. As the world's major reserve currency, the United States has an absolute say in the international trade settlement and the international financial market, just like the "hegemon" of the financial world. The Feds arbitrary monetary policy changes will trigger global inflation or abnormal capital flows, like a financial storm sweeping through the economies, while other countries suffer passively. The ripple effects of these policies are felt from the bustling streets of Tokyo to the bustling markets of Mumbai. When the Federal Reserve decides to raise interest rates, it can cause a domino effect, leading to capital flight from emerging markets and economic instability.

The decision-making mechanism of the World Bank and the International Monetary Fund is also controlled by developed countries such as the United States, and the loan conditions, like "ropes" around the neck of developing countries, force them to give up part of their economic sovereignty in exchange for limited financial support, making the road to development more difficult. These institutions, ostensibly created to foster global economic stability and growth, often impose stringent conditionalities that require nations to implement austerity measures, privatize state assets, and open their markets to foreign investment, often at the expense of social welfare and long-term economic health.

Politically, the United States frequently uses the veto power of the UN Security Council to serve its hegemonic acts, and blocks some resolutions that serve international morality but harm its interests, like overbearing "roadblocks" and hinder the pursuit of international justice. The veto power, a relic of the post-World War II era, has been wielded by the US on numerous occasions to protect its geopolitical interests, sometimes at the cost of global peace and security. For instance, in matters concerning the Israeli-Palestinian conflict, the US has often used its veto to shield Israel from resolutions that might criticize its actions, despite widespread international condemnation.

At the same time, a series of military alliance systems led by the United States, such as NATO, under the guise of "collective security", have actively intervened in various regions around the world, sometimes exacerbating conflicts rather than resolving them. These alliances, while ostensibly formed to counter common threats, often serve as tools for extending US influence and projecting power globally. For example, NATO's expansion eastward has been a point of contention, viewed by some as a move to encircle Russia and destabilize the region, rather than a genuine effort to ensure European security.

Chapter 6
The Intrinsic Value Root of "World Change" and the Evolution of Human Civilization

6.1 Value Awakening

Throughout the expansive and illustrious trajectory of human societal progress, the realization of value emerges akin to the initial beam of sunrise piercing through morning mist, charting a path for the march of civilization and infusing it with a consistent source of inherent vitality. This enlightenment is not a spontaneous revelation but is meticulously cultivated over time through the intricate interplay and amalgamation of various profound historical elements, maturing through years of refinement and trial, before finally breaking through the surface. As we turn our attention to the unfolding narrative of trade globalization and the advent of an era characterized by a "value community," the landscape of this awakening grows ever more distinct, laying bare the profound stirrings within humanity, their quest for meaning, the pursuit of value orientation, and the emergence of a new interdependent consciousness and transcendence.

As the sun's rays gently touch the horizon, they symbolize the dawn of a new era in human civilization, where the concept of value is not merely a byproduct of economic exchange but a beacon guiding our collective journey. This journey is one that has been shaped by the hands of countless individuals, each contributing their unique insights and experiences to the grand tapestry of human history. Philosophers, economists, and leaders have all played their part in weaving the threads of value into the fabric of society, each thread representing a different aspect of what we hold dear—be it freedom, justice, or prosperity.

The enlightenment of value realization is akin to the slow and steady growth of a mighty oak from a tiny acorn. Just as the oak requires nurturing soil, water, and sunlight to flourish, so too does the concept of value need the nourishment of cultural exchange, technological advancement, and philosophical inquiry. Over centuries, these elements have intertwined, creating a robust framework that supports the growth of societies and the evolution of value systems.

L. Yan, *Value-Based Civilization*, Understanding China,
https://doi.org/10.1007/978-981-96-9490-7_6

Trade globalization, a phenomenon that has accelerated in recent decades, is a testament to humanity's relentless pursuit of connection and mutual benefit. It is a manifestation of our desire to transcend geographical and cultural boundaries, to weave a global tapestry where the threads of diverse cultures and economies are interlaced. This interconnectedness has given rise to a "value community," a collective of individuals and nations that recognize the intrinsic worth of collaboration and shared prosperity.

Within this community, the quest for meaning takes on new dimensions. No longer confined to the pursuit of individual gain, it now encompasses the well-being of the global collective. The pursuit of value orientation becomes a shared endeavor, where the success of one is seen as the success of all. This shift in perspective fosters a new interdependent consciousness, one that acknowledges our shared humanity and the importance of working together to address the challenges that face our world.

As we navigate this era of globalization, we are witnessing the transcendence of old paradigms. The walls that once divided us are crumbling, making way for a more inclusive and equitable world. The value community is not just an economic construct but a philosophical one, grounded in the belief that the true measure of our progress is the extent to which we uplift one another.

In this narrative of progress, we find stories of innovation and resilience, of individuals and communities coming together to create a better future. From the bustling markets of Asia to the tech hubs of Silicon Valley, from the ancient cities of Europe to the emerging economies of Africa, the value community is a mosaic of diverse voices and perspectives, each contributing to the richness of the whole.

As we continue to chart this course, we must remain mindful of the lessons of the past, the struggles and triumphs that have brought us to this point. We must also look to the future with hope and determination, knowing that the realization of value is an ongoing journey, one that requires our collective effort and unwavering commitment to the principles of fairness, cooperation, and mutual respect. In this way, we can ensure that the light of value realization continues to grow stronger, illuminating the path forward for generations to come.

6.1.1 The Evolution of the Globalization of Trade

Trade globalization, as an epoch-making initiative in the history of human economic exchanges and interaction, is like a magnificent and surging history, carrying the collision, integration and fluctuation of countless civilizations, and has deeply rewritten the development trajectory of human society and become one of the key cradles for the awakening of value.

Back in ancient times, the Silk Road was undoubtedly a bright example of the prototype of early trade globalization. This trade passage, winding through the vast sea of sand and mountains, is like a tenacious link, closely connecting the ancient and mysterious Chinese civilization in the East with the diverse foreign civilizations in the West. At that time, China's exquisite silk, exquisite porcelain, fragrant tea and other specialties, like cultural messengers, attracted the envious eyes of people along the route of the rare spices, bright jewelry, vigorous and splendid horses, added different customs and colors to the eastern world. Behind this material exchange, a deeper level is the quiet exchange and infiltration of different cultures, ideas and skills. Businessmen, scholars and missionaries from various countries shuttled through the country, they not only brought a wide range of commodities, but also spread their religious beliefs, scientific knowledge, artistic aesthetics and political system ideas. For example, Buddhism was spread to China from India via the Silk Road, took root, spreading branches and leaves, and integrated with local culture. It has profoundly influenced Chinese philosophy, literature and art and even folk customs, and given birth to unique Buddhism Teaching cultural landscape; at the same time, Chinese paper-making, printing, gunpowder and other great inventions into the west, like a prairie fire ignited the European Renaissance and modern scientific and technological revolution, for the western world to break the dark shackles and transition towards modernization. This kind of deep interaction and communication across regions and civilizations makes people initially appreciate the diversity and interdependence of the world. Under the commercial appearance of barter and exchange, the curiosity, respect and tolerance for foreign cultures are quietly generated, laying the first seed of hope for the seeds of human value awakening.

In modern times, with the dawn of the great navigation era, the process of world history accelerated at an unprecedented pace, resembling a powerful, high-speed engine propelling humanity forward. The globalization of trade heralded a new epoch of rapid development, where the brave and fearless navigators embarked on voyages across the vast oceans, brimming with the unknown and peril. They broke through the previous geographical barriers that had constrained human communication and interaction.

Equipped with advanced navigation technology and backed by formidable military might, the European powers embarked on a rapid journey of global colonial expansion and overseas trade plunder. This period saw an influx of wealth from the Americas, Africa, Asia, and other regions into Europe, amassing substantial primitive capital that fueled the burgeoning capitalist boom. The industrial revolution gained momentum, dramatically transforming the economic and social landscapes not only in Europe but across the globe.

However, this era of expansion was not without its dark side. It brought with it bloodshed and suffering, the destruction of indigenous civilizations, the loss of countless precious cultural artifacts, the horror of the transatlantic slave trade, and the degradation of the ecological environment. Amidst this cruel historical backdrop, there emerged a complex dialectic of value awakening.

The colonial trade system necessitated unprecedented levels of contact and collision between peoples from different continents and races. Despite the inherent oppression and resistance, it also facilitated the spread of Western technology, political ideas, culture, and art across wider regions. This exchange inspired the colonized peoples to deeply reflect on their own destinies and to pursue with unwavering determination national independence and self-improvement.

From the spectacular independent liberation movements in Latin America to the Asian countries that resisted the aggression of foreign powers and sought their own paths, the spirit of nationalism and self-determination spread like wildfire. The cries for freedom and justice resonated across continents, igniting a global struggle for human dignity and rights. This period of history, marked by both exploitation and enlightenment, laid the groundwork for the modern world, shaping the geopolitical landscape and the cultural mosaic that defines our global community today.

In modern times, trade globalization has evolved to an unprecedented depth and breadth. The establishment of the World Trade Organization (WTO) is like building a fair, just and orderly "supermarket" for global trade. All countries follow a unified system of rules and participate in international trade freely and equally. Multinational corporations, like "giants" in the economic field, optimize the allocation of resources on a global scale, build a huge and complex industrial chain and supply chain system, and closely weave all countries in the world into the big network of global economic integration. Today, the birth of a smart phone is the vivid microcosm of the global production network, its parts may come from dozens of countries around the world, research and development design team brings together the wisdom of the world, production assembly and distributed to labor cost advantage regions, the finished product sold all over the world, become an indispensable part of people's daily life. In this highly integrated trade ecology, people are increasingly clear and deeply aware that countries around the world have become a community with a shared future of solidarity and lips and teeth. The economic fluctuations and trade policy adjustments in any country or region are like a huge rock on the calm lake, causing ripples and quickly spreading to every corner of the world. The us subprime mortgage crisis was triggered in 2008 The global financial tsunami swept the global economy, developed and developing countries are not immune, business collapse, unemployment, economic recession, let the world economy into the winter; and the recent trade friction, not only to both economies, but also poses a serious threat to the global industrial chain, supply chain stability, many multinational enterprises supply chain blocked, costs, the pace of global trade growth slowed down. These real crises are like a wake-up call, awakening a deep understanding of common interests and common responsibilities of mankind, prompting countries to abandon narrow unilateralist thinking, jointly explore new paths for win-win cooperation, and promote global trade in a more equitable, inclusive and sustainable direction. This process marks that, driven by the evolution of trade globalization, the value awakening of mankind has risen to a new height. We truly realize that only by solidarity, cooperation and mutual help can we move forward steadily in the stormy waves of global economic integration and create a more prosperous human civilization.

6.1.2 The "Community of Values" in the Post-Epidemic Era

When human society is immersed in the convenience and prosperity of globalization vision, the new crown outbreak like a sudden, fierce global "black storm", to destroy the corner of the world, ruthless broke the original social order and development rhythm, the human crossroads to fate crossroads, but also in the desperate produced after outbreak era unique era "value community" concept, become a powerful human value awakening into deep dimension powerful catalyst.

At the beginning of the epidemic, the virus was like a hidden, invisible but deadly "ghost". With the convenience and efficiency of modern transportation network, it spread in lightning around the world, instantly broke through the defense lines of borders, races and regions, and brought all mankind into the same life-or-death survival crisis. Whether between the skyscrapers in bustling cities or on the quiet roads in remote villages, people are facing many severe challenges such as the risk of infection, shortage of medical resources and economic shutdown. At this special historical moment, the vulnerability and interdependence of human society are fully exposed to the spotlight. Developed European and American countries, with top medical technology and strong economic strength, but in the early stage of the epidemic due to poor prevention and control, medical run, critical medical supplies such as masks and ventilators, and rising death cases in many developing countries, faced with weak infrastructure and imperfect public health system, under the impact of the epidemic, people's lives and health, economic development, and poverty has deteriorated sharply. This global disaster tragedy, let everyone really feel, human beings in the same "global village", in the face of the common enemy—virus, no country, a nation can be unique Body, each others fate is closely linked, sad and sad.

In order to cope with the challenge of the epidemic, countries around the world have launched unprecedented cooperative efforts to fight the epidemic. Although the process is full of twists and turns, it is in the practice of fighting the epidemic that the prototype of the "community of values" has gradually emerged. On the one hand, the medical research field has opened up the "acceleration" mode of global collaborative research. Scientists from various countries have broken national boundaries and exchanged research results, shared viral gene sequence information day and night through online video conferences and data sharing platforms, and jointly developed vaccines and therapeutic drugs. China was the first to isolate the novel coronavirus strain and share it with the World Health Organization, laying an important foundation for the global anti-epidemic research. Later, China, America and Europe cooperated closely, and a number of COVID-19 vaccines were released in just 1 year, creating a miracle in the history of medical research and development. This process shows that mankind can abandon prejudice and competition in the face of a major crisis of life and death, pool global wisdom and strength, and pursue the lofty value goal of the health and well-being of all mankind. On the other hand, international mutual assistance also plays a key role in the supply of medical supplies and the sharing of experience in fighting the epidemic. "Mask diplomacy" and "vaccine diplomacy" have become new hot words in international relations in a special period, especially in various countries Help, or donate supplies, or send medical experts to help the worst-hit areas. China has provided hundreds of millions of masks, protective suits, testing reagents and other anti-epidemic materials to many countries, and sent several medical teams to Asia, Africa and Latin America to help them improve their prevention and control capabilities. Russia, Cuba and other countries have also actively developed their medical expertise to provide professional assistance to other countries. Such selfless mutual assistance action across national borders has not only saved countless lives, but also conveyed warmth and hope around the world. It makes people deeply realize that mutual support and overcoming difficulties together in the face of disasters is the most shining brilliance of humanity and the fundamental place for the continuation of human civilization.

In the post-epidemic era, as the global epidemic gradually eases and enters the stage of normal prevention and control, peoples understanding and practical exploration of the connotation of the "community of values" have not stopped there, but have expanded to a deeper and broader field. In the area of economic recovery, countries are increasingly aware that it cannot achieve sustainable development, and must work together to rebuild the stability and resilience of global industrial and supply chains. For example, many countries in Asia actively promote the implementation and implementation of the Regional Comprehensive Economic Partnership Agreement (RCEP), strengthen regional economic integration, reduce trade barriers, promote regional trade and investment liberalization and facilitation, and combat the risk of economic recession. With the epidemic, the G20 (G20) and other international organizations also hold frequent summits, coordinate macroeconomic policies, reach consensus on the global economic restart and green development, and indicate the direction for the world economy to get back on track. In the field of public health, countries deeply learn lessons from the outbreak, to strengthen the global public health system construction cooperation, to the world health organization and other international health institutions of financial support and resource input, jointly improve the global epidemic monitoring and early warning, emergency response, medical supplies reserve key mechanism, is committed to improve the world The overall ability to deal with health emergencies, and to protect the life and health defense line of all mankind. In addition, in climate change, digital governance, food security, and many other global challenges, "value community concept" also widely used and depth of practice, countries have collaborative strategy, the domestic development into the global sustainable development, take concrete actions to interpret the common value pursuit of human belief and high attitude hand in hand.

To sum up, whether it is the long evolution of trade globalization, or the hasty but powerful construction of the "value community" forced by the crisis in the post-epidemic era, both are deeply intertwined and mutually reinforcing, becoming the core driving force in the process of human value awakening. They are like the shining beacon lights on both sides of the long history, guiding mankind to constantly calibrate the direction of civilization in the changing tide of world change, forge ahead towards a more fair, harmonious and prosperous future, and continue to write a magnificent epic of the pursuit of human value and civilization progress.

6.2 Value Conflict

When the human society resolutely sets foot on the long road of value remodeling and moves forward with the vision of the future, it feels that the road ahead is full of difficulties and is trapped in a series of complicated and difficult value conflicts. These conflicts are like dangerous reefs hidden in the dark of the waterway of civilization and progress, which always pose great threats. With their destructive power that cannot be underestimated, the social giant ship carrying the hopes and dreams of human beings will make turbulence violently during the voyage, and it is difficult to move forward smoothly. They like a closely intertwined, intricate network, extensive and deeply penetrated into the economic, social, cultural and international relations, and many other key areas, from the foundation violently shake the world change trajectory, to the future prospect of human civilization with layers of thick haze, deep and lasting around the pace of human forward.

6.2.1 The Contradiction Between Economic Interests and Ecological Protection

In this era of rapid development, the contradiction between economic growth and ecological protection is becoming more and more acute, and has become one of the most focal focus in the field of value conflict. On the one hand, the hot and urgent demand of economic development, like a surging and overwhelming powerful torrent, closely leads the decision-making trend of governments and enterprises of various countries in a very powerful way. In order to pursue higher gross domestic product (GDP) and achieve more abundant profit returns, all parties spared all costs to expand the production scale and intensify the development and mining of natural resources. Traditional industries, especially heavy and chemical industries, still occupy a dominant position in the economic map of many countries, just like a towering giant tree with deep roots, which is difficult to shake its solid foundation in the short term. They like to natural resources have endless greed gluttonous beast, for coal, oil, ore and other precious resources are deep and stubborn dependence, large-scale mining operations and unrestrained consumption, makes the earth's resources treasure dried up quickly, into unsustainable, also as opened the Pandoras box, caused a series of shocking, far-reaching ecological environment problems. Air pollution is like a layer of airtight haze, Heavy over the sky, the blue sky and white clouds become extravagant hope; water pollution makes the clear rivers, lakes and seas become smelly, aquatic life lost habitat; the soil degradation makes the fertile land gradually lose vitality, crop yield reduced, the balance of the ecosystem is ruthlessly broken, crumbling, on the verge of collapse. Take some developing countries as an example, in the process of industrialization and urbanization and rapid progress, in order to achieve leapfrog development in a short period of time, a large number of factories have sprung up like mushrooms, and infrastructure construction is also carried out in full swing.

However, trapped in the shortage of funds, backward technology bottleneck, environmental protection technology research and development and application lag, cash, the industrial production in the process of a large number of waste water without any effective purification treatment, was directly into the rivers, the smoke with pungent harmful waste gas drift in the air, the surrounding areas of the ecological environment, fragile ecosystem, residents health is unprecedented serious threat, all kinds of difficult diseases, the quality of life plummeted, in hot water.

As the global ecological crisis, on the other hand, the alarm sounded, ecological protection of justice like surging tide, a wave of wave, rising, become a powerful binding time, sound urged the human must reflect, decisive abandon the old extensive development model, for future generations carefully protect the blue planet, retained a pure land home for sustainable survival. In recent years, various natural disasters caused by global climate change more frequent and devastating, rainstorms, floods, heat, drought, hurricane, tsunami ravaged the world, countless homes and lives lost in the disaster; biodiversity is painful, a large number of rare species are endangered, the key link of the chain, like dominoes, once down, will trigger a chain reaction, endanger the stability of the entire ecosystem. In the face of such a serious urgent situation, the international community together, reached a series of far-reaching significance, about human fate of environmental protection consensus and agreement, the Paris agreement is a landmark important achievements, according to the agreement are solemnly to set emissions targets, ambitious for solar energy, wind energy, hydropower and other renewable energy investment research and development. We will spare no effort to boost the clean energy industry. However, there is always a huge gap between the ideal and the reality, and these seemingly promising environmental protection measures have encountered many difficulties and obstacles in the implementation process, and they are difficult to break away from. From the micro perspective of the enterprise, using clean energy often means to purchase expensive advanced equipment, upgrade the traditional production process, which undoubtedly requires huge amount of money, in the short term will be compressed profit space, and even may lead to some weak risk resistance ability of enterprise competitiveness in the fierce market competition, facing the severe test of survival, it is no wonder that many enterprises to environmental requirements, compromise, surface to meet but delay. From the macro national level, the economic development level of various countries in the world is uneven, with different development stages, and there are significant differences in interest demands. Some developing countries confidently point out that in the process of industrialization in the past hundreds of years, developed countries have long accumulated wealth at the expense of the environment, caused indelible damage to the global ecological environment, now should bear more responsibility, whether it is financial assistance, technology transfer, and emission reduction obligation sharing, while developed countries are concerned about development In the white-hot stage of economic catch-up, the country relaxed environmental standards for the sake of short-term economic benefits, thus forming an unfair competition situation and putting itself at a disadvantage. Both sides on many key issues, each other, stalemate, makes the global environmental action like fragmented, difficult to concentrate strong force, economic interests and ecological protection of the contradiction into the disturbing deadlock, become the human society to crack difficult problem, as if hanging in the sword of Damocles, warning we must find the broken.

6.2.2 *Crisis of Trust and Confidence*

Trust, this as social harmonious and stable foundation of key elements, in today's changing, challenging times, but in an unprecedented crisis, its negative impact like a raging storm, to destroy potential widely spread to political, economic, social and other key areas, to the vigorous development of human society process with heavy haze, bring many difficult problems and negative impact, let originally hopeful road become bumpy.

In the political world, public trust in the government is on a worrying decline, like a once solid building of trust, now with loose cornerstones and cracked walls. On the one hand, the corrupt acts of some officials have been repeatedly exposed in the public eye, like a bombshell, instantly exploding anger in the hearts of the people, greatly damaging the glorious image of the government as honest and impartial in the hearts of the people. These corrupt officials ignore she holds, wanton abuse of power, the public resources as personal gain tools, through manipulation, power cheap means, for individuals or small groups for improper interests, enrich themselves, make people to the governments integrity and impartiality, trust cracks, like a difficult to heal the wound, gradually eroding the trust link between the government and the people. On the other hand, some government decision-making process lacks due transparency, and the timeliness and comprehensiveness of information disclosure is seriously insufficient. It is like building an invisible but solid barrier between the government and the public, causing unprovoked suspicion and dissatisfaction of the public. For example, in the major decision-making process of urban planning, which concerns the future development layout of cities and the quality of life of the people, if the people are not widely solicited. It means to fully listen to the voice of the people, and explain the decision basis, potential impact and the balance between the advantages and disadvantages in detail. People often tend to assume that there is an shady operation in the government decisions, which leads to a great reduction in the trust in the government. In the long run, this lack of trust will not only like chronic poison erode the government's credibility, more likely to cause people in the policy implementation link, refused to cooperate, make social governance effectiveness, even in extreme cases, like a powder keg lit, cause social unrest, a serious threat to social stability, let the whole society into chaos and disorder.

In the economic field, the crisis of confidence is like a fierce plague in the business field, spreading everywhere, and its negative effects are mainly reflected in all aspects of business activities. With the wheel of the market economy rolling forward, rapid development, the phenomenon of commercial fraud is like weeds in the commercial soil crazy growth, emerging in an endless stream. Some unscrupulous enterprises, in spite of business ethics and integrity principles, adopt false publicity to exaggerate the efficacy of products, mislead consumers, and play their trust in the production process, shoddy, seriously infringe the legitimate rights and interests of consumers, and discard the commercial reputation. In recent years, food safety problems frequently, repeatedly hurt the public sensitive nerve, some food enterprises ignore consumer life and health, crazy use inferior raw materials, illegally add preservatives, melamine and other harmful substances, make the public to the food market trust, talk about "food" color change, every eating is full of concerns; in the financial sector, illegal fund-raising, fraud and other illegal and criminal activities, like a hidden in the dark of the fatal trap, always threatening the property safety of the public. Some criminals are investing with high returns Under the guise, they made up various false projects, persuade people to invest money, once they get the money, disappear without a trace, disrupting the financial order, causing investor's confidence in the financial market, like a panic, and have doubts about any investment opportunity. In addition, the trust relationship between enterprises also become increasingly vulnerable under the impact of the interests of, contract default, business spy, if the business cooperation link on the thorns, greatly increased the transaction cost, hindered the normal circulation of economic factors and healthy development, and makes the vitality of the market economy is difficult to fully release, as a giant bound hands and feet, unable to display.

In the social field, the warm and close trust bond between people has become increasingly weak under the impact of multiple factors such as the fast-paced lifestyle of modern society, high competitive pressure and large-scale population flow brought by the wave of urbanization. Like a gauze full of holes, it may be broken and dissipated at any time. Neighborhood relations are gradually alienated in the city of high-rise buildings, and the close ties of traditional communities are ruthlessly broken, replaced by closed security doors and cold eyes, as if everyone had closed themselves in a small island. In daily life, people become more and more vigilant when facing strangers. Like a frightened bird, they are often cautious and hesitant to seek others help for fear of getting into trouble and taking unnecessary risks. The problem of "help or not" is a typical epitome of this trust dilemma. When an old man accidentally falls to the ground, passers-by lend a helping hand, their heart is often full of entanglement and worry, fear of being blackmailed and suffering from any disaster, so that they dare not come forward to help him. This phenomenon not only profoundly reflects the serious lack of trust between individuals, but also is like a mirror, mercilessly revealing the fragility of the whole social moral trust system, which is painful, as if telling the indifference of the society cannot help but. Moreover, the virtuality and anonymity of network social networking should be a beneficial platform to expand interpersonal communication and enrich social experience, but to some extent, it has become the "accomplice" of the trust crisis, aggravating the trust dilemma. Online fraud and the spread of false information make peoples communication in the virtual space as if they are in the fog, full of doubts and uneasiness. It is difficult to establish a real, reliable and honest interpersonal relationship, further eroding the foundation of social trust and making people lose their security in the virtual world.

6.2.3 Information Island

The mighty tide of the information age should be like a rain to moisten all things, making knowledge and information like unimpeded rivers, nourishing every corner of human society, and becoming a powerful engine to promote social development and civilization progress. Reality, however, is a blow, many “information island” such as stubborn rocks, become another outstanding problems, they like invisible but indestructible barrier, seriously hindered the efficient circulation of information and sharing, to the operation of all levels of social inconvenience and heavy losses, if to the wheels of social forward on the heavy shackles.

At the enterprise level, due to the cruel pressure of business competition, many concerns about data security and many other factors, different enterprises have done their own business and tried their best to build an independent and closed information system. These systems are like an isolated “information fortress”, the lack of compatibility and connectivity between each other, the exchange of information is abruptly blocked, if separated by a high wall. Even within large enterprises, there are also towering information barriers between departments, like mountains, the collaboration between departments becomes extremely difficult. Take the marketing department and the RESEARCH and development department as an example, the marketing department is at the forefront of the market, dealing with customers all day, and mastering the massive and first-hand customer demand information, which is crucial for the research and development department to accurately locate the product direction and optimize the product function, just like a beacon in navigation, pointing out the direction for research and development. Reality, however, is often due to internal communication, information transmission channel blockage, lead to the market department to master the precious customer demand information is difficult to smooth, timely transfer to research and development department, make the research and development and the actual market demand, eventually to the market products often not accurate fit customer expectations, failed in the fierce market competition, as lost ships, adrift in the business; Look at the financial department and operation department, the financial department holding enterprise capital flow, cost accounting and other key data, the operation department is familiar with the production process, supply chain management and other practical links, but the data of the two like two parallel worlds, fragmented, make the enterprise management like fog, difficult to fully and accurately grasp the real situation of the enterprise operation, and then make strategic decisions as a blind, lack of scientific basis, easy to deviation, bring huge risks to the development of the enterprise, as if in the dark, may fall into the abyss. In the more complex scenario of cross-enterprise cooperation, the problem of information island is even worse and becomes extremely difficult. Due to different enterprise data format vary widely, interface standards, the lack of unified specification, data sharing and business synergy like a difficult “international marriage”, difficult, not only lead to project progress delays, missed market opportunities, also greatly increased the cost of cooperation, consumption enterprise precious resources, severely weakened the overall competitiveness, makes the enterprise in the tide of globalization market struggling, like upstream, not to advance.

Among government departments, the phenomenon of information island also exists widely, which is like a "information vassal" fighting for themselves, which seriously restricts the service efficiency and governance level of the government. Based on the professionalism and independence of their own business, different functional departments do their own affairs, and firmly close the massive information resources in their own departments, forming an isolated information stronghold without support. The information circulation is like falling into a pool of stagnant water, with no vitality and vitality. Taking the complex system project of urban management as an example, the transportation department holds key traffic information such as real-time traffic conditions and vehicle violations, the urban management department knows the city appearance such as urban environmental sanitation and illegal stalls, and the public security department is familiar with the core data such as social security dynamics and detailed information of registered population. Unfortunately, however, the lack of a set of effective information sharing mechanism between these departments, if each guard an information island, when a comprehensive urban problems such as large activities, such as emergency treatment, cannot quickly break the barriers, integration of the information, concerted action, lead to low management efficiency, problem solving procrastination slow, bring people daily life inconvenience and trouble, seriously affect the normal operation of the city and the harmonious development of the city, like The city's "blood vessels" are blocked and cannot run smoothly. Moreover, the information exchange between government departments, enterprises and social organizations is also like a layer of "thick barrier", not smooth enough. Government in policy, provide public services, due to the lack of enterprise actual operation situation, social organization service demand of deep communication, difficult to accurately grasp the social pain points and enterprise demands, makes the policy often pertinence is not strong, effectiveness, unable to meet the demand of social development, cause the waste of public resources and social governance lag, if the arrow deviated from the bulls eye, cannot hit.

In the field of academic research, the information island is also like a haze, to a certain extent, closely restricts the pace of scientific research progress, and restricts the enthusiasm and strength of mankind to explore the unknown field. With the increasing refinement and specialization of academic research, different disciplines and different research institutions are like building invisible walls, and the barriers have become more and more strict. On the one hand, researchers are often immersed in their own subdivided research field, just like the "frog in a deep well", with limited vision, insufficient attention to the frontier dynamics of other fields, and lack of active awareness of interdisciplinary communication and cooperation. That has left some research projects behind closed doors, The research results were repeated and inefficient, Innovative breakthroughs are, if not, out of reach, A lot of scientific resources are wasted, Delay the overall process of human research, Like a snail crawling, Forward slowly; on the other hand, Academic databases and scientific research literature resources are like "knowledge pearls" scattered around the country, Scattered in various colleges and universities, research institutes, due to copyright restrictions, access rights and other cumbersome issues, researchers in the access to comprehensive information as if in the thorn bushes trekking difficult, difficult, greatly restricting the broadening of the research horizons, hindering the inheritance of knowledge and innovation, making it difficult to stand on the shoulders of giants to look into the distance, to overcome the scientific challenges, and not to mention that human beings on the universe of everything, the laws of nature, in-depth exploration and comprehensive attack, as if locked feet walkers cannot make strides forward. In-depth exploration and comprehensive attack, as if locked feet walkers, unable to stride forward.

6.2.4 Order Imbalance

Order, just like a lighthouse standing in the vast sea, guides the direction for the normal operation of the human society, and is the basic principle to ensure the steady progress of the society. However, in today's changing situation of the world, whether the spectacular international order, or close to the livelihood of the domestic social order, are different degrees of imbalance, cause many deep, sharp value conflict, like a storm sweeping the ship of human civilization, let people into deep worry.

From the macro perspective of the international order, with the booming rise of emerging economies like bamboo shoots after a spring rain, the old international political and economic order dominated by the western developed countries looked like a building with a shaken foundation, which was strongly impacted and crumbling. In the collision of the new and the old, the new balance has not yet been established, and the international order has fallen into a period of transition, like a ship losing its helmsman in a stormy wave, at a loss and drifting with the tide. At the political level, hegemonism and power politics the two cancer still stubbornly parasitic on the international political body, individual powers with its strong military and economic strength, trample on the basic principles of international relations, if waving the stick of hegemony, unscrupulous interference in domestic affairs, the international rules to "use, not abandon" double standards, the international stage as their backyard, do whatever you want, seriously damage to the fairness and justice of the international order. In the Middle East is a typical portrayal of the tragedy, long-term hegemonism, war, innocent people displaced, life and property, suffered huge losses, regional development stalled, the dawn of peace and development by war smoke, if the darkness of the earth, can't see a glimmer of hope. In the economic field, the global economic governance system is like an old house in disrepair, with many irrationality, and urgently needs to be repaired. The decision-making mechanism of the international economic organizations such as the International Monetary Fund and the World Bank is like a tilted balance, preferring the developed countries and developing them.

In the map of the international economic order, just like the vulnerable group, in the process of making international economic and trade rules, they can only passively accept the established rules, and their reasonable demands are blocked by an invisible barrier, and it is difficult to express and extend smoothly. As a result, the imbalance of global economic development is becoming more severe, and the gap between the rich and the poor is like a deep and bottomless barrier, clearly dividing the world into two poles. Developed countries enjoy prosperity rich, while developing countries are struggling in poverty and development difficulties, the north-south contradictions is increasingly sharp, like a cold sword of damocles, hanging above the global economic development, time threaten the prosperity and stability of the world economy, let the path of the global economy is full of haze.

Focusing on the level of domestic social order, the phenomenon of social class solidification is becoming more and more prominent. Like a layer of cold and hard frost, it ruthlessly seals the due vitality source of social flow and becomes the key fuse to induce the imbalance of order. The flow channel between different classes, if the years of silt gradually silt blocked, become more narrow and narrow. The people at the bottom, with fervent dreams, hope to climb up step by step, break free of the shackles of fate, and achieve class leap. However, the reality is like a cold wall, standing in front, they feel that the road ahead, difficulties, far beyond imagination. The uneven distribution of educational resources is undoubtedly a sharp edge, which brutally cuts off the hope of poor areas and vulnerable groups to change their fate. High-quality education resources are firmly monopolized by a few advantaged areas and classes, in sharp contrast, children in poor areas can only be confined in the shabby and dilapidated classroom, holding old, musty books, eyes full of desire for knowledge, but limited. However, children in developed areas can leisurely enjoy the audio-visual feast brought by modern multimedia teaching, bathed in the careful guidance of international teachers. The two sides in the starting line of life has already opened a huge gap. Such disparity of starting point difference makes the poor children in the subsequent job market competition, it is difficult to emerge, the dream of the realization of the unfair distribution of wealth, like a dark circle, ruthlessly involving the bottom of the people, the deeper and deeper, difficult to extricate themselves. A few rich class at the top of the wealth pyramid, with huge wealth, extravagant life, spending money, enjoy the world; while the vast number of middle and low income groups under the pressure of life, difficult to breathe, income growth slow as a snail, the cost of living is like rocket climbing, housing, health, education these three towering mountains heavy pressure on their shoulders, let them suffer. In the long run, the class curing caused by the social discontent, like an internal lava, the violent eruption of the volcano, if not properly guide and resolve, is likely to induce social contradictions intensified, frequent mass incidents, like stones into calm lake, instant ripples of social order, violently impact the social stability and harmony, the normal social order is mixed The chaos has brought heavy disaster to the lives of the broad masses of the people.

In addition, under the current digital wave, cyberspace has been deeply integrated into people's daily life and become an indispensable "second space". But what is worrying is that the imbalance of order in cyberspace cannot be underestimated, like a "wild west" in a virtual world, full of chaos and disorder. As the Internet as an invisible but ubiquitous spider web, with astonishing speed quickly spread, its virtual and anonymity characteristics, if for some malicious people with a layer of hidden cloak, let them instantly bullying, wanton, if runaway, rampage in the network world, do whatever you want. They release all kinds of illegal, false and harmful information, these information like cancer, wanton spread, mislead the public opinion, the network public opinion field, seriously disrupt the network order; what is more, unscrupulous network violence, with vicious, unbearable words to the victims, the victims of the victims in the virtual world of the moral and legal red line, wanton violation of others privacy, others personal information without reservation, completely not to the dignity and rights of others. These chaotic behavior, not only brutally destroyed the network space should have harmonious atmosphere, more like a virus, quietly penetrated into the real society, the peoples values and moral bottom line launched fierce erosion, make the network this piece should be convenient information exchange of pure land, gradually become a hotbed of bad behavior wanton breeding, to all parties hand in hand, intensify regulation, reshape the network space clear order, also a net network world.

To sum up, the contradictions between economic interests and ecological protection, such as trust crisis, information island and order imbalance, are just like a grim and terrible gap on the development of human society, and become heavy shackles that hinder the progress of civilization. If not timely and effective solutions to resolve these conflicts, the great ship of human society, carrying countless hopes and dreams, will be uncertain under the ravage of the wind and waves, and even very likely to hit the rocks and sink. Therefore, exploring practical solutions and reshaping the harmonious and stable value system has become an urgent and urgent task for all mankind in the current era, which is related to the future destiny of mankind and the rise and fall of its well-being.

6.2.5 Global Divide

In today's soaring under the tide of the era of globalization, "global gap" this thorny problem more prominently exposed, like a deep and difficult to cross the huge cracks, prominently stood between countries around the world, from many key dimensions mercilessly tore the pace of human society as a whole, to the world to collaborative journey together with a layer of thick almost depressed haze.

From a detailed analysis of the economic dimension, the economic gap between the north and the south is like continuing to be pulled by an invisible hand, showing a growing trend, like a gap of accelerating expansion, making the world polarization situation more and more shocking. On the one hand, the developed countries rely on deep solid industrial accumulation, always walk in the forefront of science and technology innovation hard power, and the dominant voice in the international financial system, as a dominant chess, firmly control the key link at the top of the global industrial chain and supply chain, easily to high profits. Take the information technology industry, a few industry giants of the developed countries, like a heavy overlord, firmly monopolize the core technology and key patents, with technology authorization, dumping, almost greedily earned most of the global profits in the field, let wealth like a surging tide, continuously, unimpeded into the country. On the other hand, the majority of developing countries, most of them can only be at the bottom of the industrial chain, relying on the export of cheap labor and the external sale of primary resources, they are struggling to maintain the weak flame of economic growth. Not only do these countries get little added, but they work hard on poor soil but gain little Farmers, who are vulnerable to price fluctuations in international markets and victims of trade protectionism, enjoy the stability and sustainability of economic development like a candle in the wind. Such a huge economic gap directly leads to developing countries weak investment in infrastructure construction, education and medical care and other key areas of peoples livelihood, and further restricts their own development potential. The gap with developed countries has snowballed, trapped in a vicious circle, which is difficult to recover.

On the other hand, inequality in the financial sector is equally eye-dropping. The international monetary fund, the world bank and other core hub in the field of international financial institutions, its decision-making power is long been tightly in the hands of developed countries, developing countries in eager to obtain loans, need funding critical moment, often have to face harsh conditions, was forced to accept the additional, sometimes even completely do not fit their own development stage and the characteristics of the economic reform terms. These overlord terms, like a hidden time bomb, may at any time in the economic system of developing countries, unrest, makes the developing countries in the international financial game without smoke increasingly passive beaten battlefield, difficult to use the power of financial leverage to achieve leapfrog development, can only watch and the financial gap between developed countries, helpless.

From the perspective of science and technology, the global technology gap is widening at a worrying rate, like a seemingly invisible but solid barrier that is brutally hindering the balanced dissemination and diffusion of knowledge and cutting-edge technology on a global scale. Developed countries invested heavily in scientific research innovation of the vast ocean, with preferential treatment, top scientific research environment, brought together the world's most outstanding research talents, carefully create a world-class level of scientific research infrastructure, and in artificial intelligence, biological medicine, aerospace and other cutting-edge technology, patent achievements as the fruits of harvest branches, constantly expanding the boundary of human cognition, leading the surge of global science and technology development. In contrast, developing countries, due to the long shortage of scientific research, as dry land desire for showers, scientific talent drain, if the injured, research atmosphere and weak as fog, many difficulties intertwined, causing many key technology research and development lag, difficult to break through the core technology bottleneck, in the competition of science and technology track can only reach, at a disadvantage. This not only makes it difficult for developing countries to develop high-end manufacturing and emerging industries, but also pays a huge price for every further development, which means that they have a voice in global science and technology Competition for without confidence, like soldiers without weapons, difficult to participate in the future rules of science and technology, can only follow behind the developed countries, but into "technology introduction-backward-introduction" passive endless cycle, the heavy shackles of technology gap, tightly locked the throat of their development.

Carefully observed from the social and cultural dimension, the global cultural gap is also a problem that cannot be underestimated, like a hidden reef hidden in the depths of human society. Western culture, especially in Hollywood movies, pop music, fashion trends as the typical representative of American culture, with its strong to almost pervasive media machine and business promotion network, spread worldwide, like a surging cultural tsunami, occupy the mainstream culture market, quietly but penetration to influence the values of the people around the world values, way of life and aesthetic standards. This almost one-way cultural output frenzy, to a certain extent, the inheritance of local culture and development caused a violent impact, many ancient and unique charm, bearing the memory of cultural treasures, and regional characteristics of niche culture, are facing the endangered crisis, cultural diversity of the bright pearl of human civilization, is gradually eroded, lose the past luster. At the same time, due to the differences in different historical precipitation, religious belief, social system and other factors, there are many obstacles in cultural understanding and communication, just like a thick fog, which is easy to cause unnecessary mistakes The solution and conflict have further deepened the global cultural gap, making the human society in the road of seeking deep cultural consensus, and added a lot of invisible but thorny obstacles for global cooperation.

6.2.6 World Change

Face to face with the above complex, like the value of the conflict and a series of global issues, the world change has become era in the development of imminent intense demands, a comprehensive, deep, disruptive power of global change is surging momentum, carrying the dilemma shackles of all mankind, to new civilization height climb high expectations.

In the political field, the international political landscape is like a great ship sailing in the tide of The Times, and it is at a critical turning point of profound adjustment. Emerging economies are like a group of vibrant and thriving rising stars. Many organizations, represented by BRICS countries, are growing in international influence like the rising sun. They clearly and actively advocate multilateralism, raise ones arm and shout, Call for a new international political order that is more fair, just and inclusive, No longer willing to be as in the past, In a supporting role on the edge of the international arena, But with a confident, courageous and active new attitude, Quietly playing a supporting role on the grand game of global governance, On important international platforms such as the United Nations and the G 20, Speak the sword, In order to promote the democratization of international relations, Trying to break the long-held hegemonic monopoly pattern of traditional powers with great power, Make the world political stage more diverse and balanced, Like building a solid bridge, To provide a solid and reliable political guarantee for global stable development.

The economy is also a surging wave of change. On the one hand, the digital economy is like a runaway dark horse, with big data, a series of emerging technologies such as cloud computing, block chain as a strong driving force, like a magical painter, reshaping the global economic map. Traditional industries have accelerated the digital transformation process, intelligent manufacturing, online services and other new business models are booming, greatly improve the production efficiency and resource allocation efficiency, as injected a shot in the arm to economic development; cross-border e-commerce has broken the space-time constraints of trade, like opened up a new commercial highway, let small and medium-sized enterprises can also set foot on the big stage of global trade, to broaden the international market, continuously inject fresh new power for global economic growth. On the other hand, the concept of green economy is like a seed that has taken root and sprouted in people's hearts. As time goes by, it has been deeply rooted in the development strategies of countries around the world and deeply rooted in people's hearts. With governments setting ambitious carbon reduction targets, solar and wind power costs are falling as much as technology advances and applies on a large scale, following a solemn pledge to increase research and investment in renewable energy waterfall, continue to reduce, new energy automobile industry is like running cheetah, development momentum, gradually replace the traditional fuel cars, leading the global economy to green sustainable development, in the long economic development, always protect the earth home, realize the harmonious coexistence of economic prosperity and ecological protection, draw a new picture of a beautiful home.

Changes in the social field are also progressing in an orderly way. With the development of The Times, peoples pursuit of the quality of life is increasing, and the call for social fairness and justice is also growing. A vast movement in the pursuit of social justice and tolerance of all things has been quietly set off around the world. Notable and impressive progress has been made in gender equality, more and more women are looking like broken butterflies, brave to break through the professional ceiling, emerging in various once male-dominated fields like politics, technology, and commerce, bloom out of their own bright light, to contribute to the more balanced development of society; multiculturalism is like a warm spring breeze, being widely advocated, people of different ethnic, ethnic and religious backgrounds begin to respect and embrace each other, like working together to build a harmonious and diverse big family, reduce the breeding ground of cultural conflicts and social contradictions; basic public services such as education and medical care are becoming more inclusive, using innovative means such as telemedicine and online education, like building a bridge across time and space, so that remote areas and vulnerable groups can also easily enjoy high-quality resources, improving the well-being of all the people, to narrow the social gap, to lay a solid foundation for the harmony and stability of the human society, like for a tall building to lay a deep-pile.

The field of science and technology is undoubtedly at the forefront of this global change. Scientific and technological innovation shows an explosive growth trend, like a colorful fireworks show, lit up the road of human progress. The popularity of 5G technology, Let the information transmission like lightning to achieve high speed and low latency, Opin a new era of the Internet of everything, For intelligent transportation, industrial Internet and many other fields, Make it glow with a new vitality; Gene-editing technology is like a magic key, is expected to overcome many difficult diseases, Rewrite the code of human life, to extend the healthy life span, bring the dawn to human health and well-being; quantum computing technology is like a superhero, breaking through the bottleneck of the traditional computer computing power, will revolutionize everything in cryptography and materials science, provide super tools for human beings to explore the unknown universe and solve complex problems, as if opening a door to a whole new world. Scientific and technological change is not only changing the way of human production and life, but also like a powerful engine, accelerating the evolution of human civilization, allowing us to view the world from a new perspective, unlock no as much as possible, and march toward the stars and the sea.

However, we must be soberly aware that the road to world change is not a smooth road, but full of thorns and frustrations, facing many severe challenges and strong obstacles one side, old interest groups are like a group of stubborn nest-guarding beasts, To safeguard their own vested interests, At all the means employed, Against the rolling tide of change, Trying to hinder the wheel of history; on the other hand, While technological innovation brings infinite surprise and hope, Has also spawned a series of new ethical, new risk problems, Such as the algorithmic bias in AI, Like the "colored glasses" hidden behind the intelligence, Ethical controversy over gene editing, Like the sword of Damocles hanging over the head of technology, All of this calls for global cooperation, To build consensus with wisdom, Set rigorous rules, Carefully address the challenges. However, no matter how thick the thorns ahead, how rugged the road of change, the general trend of world change has been like a surging tide, unstoppable. As long as all mankind think and work together, lead the direction with wisdom and overcome difficulties with courage, we will overcome the difficulties and difficulties, usher in a much brighter new world, realize the great rejuvenation of human civilization, and create a new glory in human history.

6.3 Value Consensus

In today's world, the process of globalization is rolling forward like a surging tide, deeply intertwined with the fate of all countries in the world, making every country unable to be immune. However, the world pattern is as complex and unpredictable as the changing chess game, new forces are rising, and the old order continues to be impacted. In such a big background, human society is like a ship in the waves, facing many difficult and complicated difficulties and challenges, from the ecological crisis of climate change, to local wars caused by humanitarian disaster, from the disparity between the rich and poor social contradictions intensified, to cultural conflict of ideological collision and barriers, all let the road of human society is full of thorns. Carefully analyzing the root causes of these problems, it is not difficult to find that, in the final analysis, they are mostly derived from the disorder and imbalance of the value system, as if the foundation of a grand building appeared loose and dislocation, making the superstructure crumbling. Exploring the road of value consensus has become the key to break through this fog and lead mankind to a bright future. It is like a towering lighthouse standing in the vast sea, unswervingly guiding the direction of mankind in the turbulent tide of The Times, illuminating the long march of working together to create a harmonious world. Take here, we will deeply and carefully analyze the core elements of value consensus construction and its far-reaching significance from multiple dimensions, and strive to find a solid theoretical support and feasible practical road for the continuous progress of human society.

6.3.1 Value Order

Value order, as the value system of the skyscraper cornerstone and pillar, its stability or not directly about human society can run orderly, like the precision clock that group of interlocking, collaborative gear, every gear accurate rotation, to ensure that every link in an orderly way, every minute. It covers the category is very wide, from People's Daily life familiar, word of mouth ethical micro principles, to regulate the behavior of the whole social norms, maintaining social stability, to coordinate the relations, the maintenance of international world peace is the macro architecture, like a invisible but tough net, through all aspects of human life, comprehensive constraints and guide the human every move.

In the intricate tapestry of human society, the value order serves as the fundamental framework that underpins the very fabric of our existence. It is akin to the intricate cogs of a precision clock, where each gear must rotate with exactitude to ensure the smooth functioning of the entire mechanism. This intricate system encompasses a vast array of elements, from the micro principles of ethical behavior that are part of our daily lives and passed down through word of mouth, to the macroscopic norms that govern the behavior of entire societies. These norms are essential for maintaining social stability and coordinating relationships, extending their influence to the maintenance of international peace and security. This invisible yet resilient net weaves through every aspect of human life, imposing comprehensive constraints and offering guidance on every human endeavor.

At the social level, a stable and reasonable value order is like a moderate and firm guardian, which can extremely effectively regulate individual behavior and escort social harmony and stability. Over time, traditional moral concepts such as respecting the old and loving the young, honesty and trustworthiness, friendliness and mutual assistance have been deeply ingrained in the hearts of people. These values have become an invisible yet solid behavioral compass, guiding individuals in their daily interactions. In the tranquil and peaceful setting of community life, neighbors consistently adhere to the principle of mutual help. When someone encounters difficulties, there is always a helping hand extended—be it in taking care of the sick elderly, ensuring they receive the warmth and care they deserve, or in coming together to repair the damaged houses, pooling the collective strength of the community. This warm and harmonious community atmosphere not only makes residents feel the strong human touch but also enhances the cohesion of the community, fostering a sense of belonging and creating a better home for all.

In the competitive business field, the integrity of businesses and the ethical conduct of entrepreneurs play a pivotal role in shaping a fair and just market environment. Companies that uphold transparency and accountability in their operations often find themselves reaping the rewards of customer loyalty and a positive reputation. For instance, a local grocery store that offers fair prices and quality products without hidden charges builds trust with its customers, who in turn become regular patrons and advocates for the business. This trust is the bedrock upon which long-term success is built.

On a larger scale, the value order is crucial in the governance of nations. Governments that adhere to principles of justice, fairness, and transparency are more likely to gain the trust and cooperation of their citizens. This, in turn, leads to more effective policy implementation and a more stable political environment. For example, a country that invests in education, healthcare, and infrastructure not only improves the quality of life for its citizens but also lays the groundwork for sustainable economic growth. The value order, in this case, acts as a guiding light, ensuring that the nation's development is inclusive and equitable.

Internationally, the value order is the cornerstone of global cooperation and peace. Nations that respect international law and uphold the principles of mutual respect and non-interference in internal affairs contribute to a more stable and predictable world order. For example, countries that engage in diplomatic negotiations and peaceful resolutions to conflicts, rather than resorting to military action, foster an environment conducive to global security and prosperity. The value order, in this context, acts as a unifying force, encouraging nations to work together towards common goals and shared benefits.

In conclusion, the value order is an indispensable element that permeates every level of human society, from the individual to the global community. It is the invisible yet resilient net that guides our actions, fosters harmony, and ensures the smooth functioning of our complex world. By upholding and promoting a stable and reasonable value order, we contribute to the creation of a more just, peaceful, and prosperous society for ourselves and future generations.

6.3.2 Value Mutual Learning

Human civilization is akin to a vast and diverse garden, where each country and nation represents a unique and vibrant flower, blossoming with its own distinctive allure and captivating fragrance. Imagine this garden spread across the continents, with the Eiffel Tower in Paris standing tall like a lily among the foliage, symbolizing France's elegance and innovation. In the heart of Africa, the Great Pyramid of Giza rises like a sunflower, representing Egypt's ancient wisdom and architectural prowess. The Statue of Liberty in New York City gleams like a radiant sunflower, embodying the spirit of freedom and democracy in the United States. These flowers, each with their own cultural and historical significance, contribute to the rich tapestry of human society, much like the vibrant colors and scents that fill a lush garden with life and beauty.

The process of mutual learning can be compared to a diligent bee, tirelessly flitting from bloom to bloom, collecting the most precious nectar from each. This nectar symbolizes the essence of different cultures, which, when combined, nourish the bee and inject a continuous flow of vitality into the progression of civilization. Just as bees pollinate flowers, spreading life and diversity, so do people exchange ideas and knowledge, fostering growth and innovation. This exchange propels human society to new heights and broader horizons, breaking down barriers of region and culture. It acts like a vigorous spring breeze, dispelling the mists of misunderstanding and promoting the exchange and integration of various outstanding values and practical experiences. Through this collision of wisdom, a spark ignites, leading to a common elevation and the blossoming of an even more dazzling brilliance.

Reflecting upon the long and winding river of history, we see that exchanges and mutual learning among civilizations have been a constant and unceasing process. This exchange is like a flowing river that carves its way through the landscape, creating numerous brilliant chapters and leaving behind a legacy of countless treasures that have propelled human development forward. The ancient Silk Road, a legendary route that traversed the vast expanse of the Eurasian continent, was not merely a conduit for commerce and trade. It was also a vibrant conveyor belt of cultural values and knowledge. Eastern silks, with their smooth, delicate textures and exquisite craftsmanship, drifted like wisps of brilliant clouds, bringing a touch of luxury and sophistication to the courts of Europe. Alongside these luxurious fabrics, spices from the East Indies infused the kitchens of the West with exotic flavors, transforming culinary traditions and sparking a gastronomic revolution.

In the realm of philosophy and science, the exchange of ideas was equally transformative. The works of Aristotle and Plato, once confined to the libraries of ancient Greece, found their way to the Arab world, where scholars like Avicenna and Averroes translated and expanded upon these texts, enriching the intellectual landscape of the Middle Ages. This cross-pollination of knowledge led to breakthroughs in medicine, mathematics, and astronomy, benefiting societies across the globe. The Renaissance in Europe was, in many ways, a flowering of this shared wisdom, as artists and thinkers rediscovered the classics and built upon the foundations laid by their predecessors in other lands.

As we continue to navigate the ever-evolving landscape of our global society, it is crucial to remember that the garden of civilization thrives on diversity and exchange. Each culture, with its unique contributions, is like a different species of flower that adds to the garden's splendor. By valuing and preserving this diversity, we ensure that the garden remains vibrant and full of life, a testament to the enduring power of mutual learning and shared progress.

6.3.3 Value Interconnection

In the wake of the digital revolution that has swept across the globe, the concept of value interconnection has emerged as an unstoppable force, akin to an invisible yet potent network that tightly binds every facet of human society. This phenomenon has enabled the transmission and sharing of values to transcend temporal and spatial barriers, ushering in a novel paradigm of interaction that has profoundly transformed the way people live, work, and communicate.

The burgeoning growth of Internet technology stands as the cornerstone of this value interconnection. It functions as an engine with extraordinary explosive potential, propelling human society into a new epoch characterized by rapid information exchange and shared value systems. Social media platforms such as Facebook and WeChat have evolved into colossal gathering spaces, attracting billions of users from across the globe and demolishing geographical barriers. This has effectively transformed the world into a "global village," where individuals can instantly disseminate their life experiences, knowledge, and cultural traditions. For instance, a heart-wrenching plea from children in remote mountainous regions, yearning for education, can now spread across the digital landscape at lightning speed. This message has the potential to capture the attention of compassionate individuals worldwide in a matter of moments, prompting widespread generosity. Consequently, funds and resources are mobilized, bridging the gap between hope and reality, ensuring that mountain children receive new textbooks and bright classrooms, thereby keeping the flame of knowledge alive.

Moreover, academic forums and online education platforms have become treasure troves of knowledge, where the pursuit of learning seems endless. Scholars and experts can now exchange cutting-edge research findings in real-time, fostering a dynamic environment where ideas storm and ignite wisdom. Experts from diverse countries and disciplines converge online to address pressing global challenges such as cancer and climate change. Students are no longer confined to the walls of a traditional classroom; they can now access the world's most prestigious courses from the comfort of their homes, listening to the teachings of renowned masters. This democratization of knowledge has not only broadened educational horizons but has also facilitated a more profound understanding and appreciation of different cultures and perspectives.

The digital age has also given rise to new forms of collaboration and innovation. Crowdfunding platforms, for example, have revolutionized the way projects are funded and supported. A creative individual with a groundbreaking idea can now present their vision to a global audience, seeking financial backing from those who share their passion. This has led to the successful launch of countless startups and initiatives that might have otherwise gone unnoticed in a pre-digital era.

Similarly, telecommuting has become a viable option for many, thanks to the advancements in digital communication tools. Professionals are no longer tethered to a specific location, enabling them to work from anywhere in the world. This flexibility has not only improved work-life balance but has also opened up opportunities for collaboration across different time zones, further enriching the global exchange of ideas and expertise.

In the realm of entertainment, streaming services have transformed how people consume media. With just a few clicks, one can access a vast library of movies, TV shows, and music from around the world, fostering a more interconnected and culturally diverse entertainment landscape. This has also given rise to a new breed of content creators who can reach international audiences without the need for traditional gatekeepers.

The value interconnection fostered by the digital age has also had a profound impact on social movements and activism. Social media has become a powerful tool for organizing and mobilizing support for various causes. A single hashtag can now unite millions in solidarity, amplifying voices that advocate for change and holding institutions accountable. This has been evident in movements such as MeToo and Black Lives Matter, where digital platforms have played a crucial role in raising awareness and driving social change.

In conclusion, the surge of the digital age has not only reshaped the way we connect and share values but has also paved the way for a more interconnected, informed, and collaborative world. The invisible yet powerful network of value interconnection continues to expand, promising even greater transformations in the future.

6.3.4 Value Exchange

Value exchange emphasizes the in-depth communication, understanding and integration of different value systems, cultural backgrounds and interests. It aims to eliminate estrangement and misunderstandings, build a bridge for empathy, and build a solid cultural foundation for building a community with a shared future for mankind, and enable people around the world to work together and unite as closely as the seeds of a pomegranate.

In a multicultural society, different ethnic groups and religious groups uphold different values, just like the stars shining. Respecting and promoting the exchange of values can resolve potential conflicts, gather social synergy, and create a harmonious and inclusive social atmosphere. Take Singapore as an example. As a country of immigrants, the Chinese, Malay, Indian and other diverse ethnic groups coexist in harmony, just like a colorful picture of ethnic integration. Through the implementation of multicultural policies, the government takes good care of the cultural flowers of every nation. Encouraging all ethnic groups to retain their distinctive cultures, Understand each other's values behind festivals and customs, For example, during the Chinese Spring Festival, The street is decorated with lights, Dragon dance and lion dance is very lively, The Malay and Indian compatriots are all involved, Feel the festive atmosphere of traditional Chinese culture, Taste traditional delicacies such as dumplings and rice cakes, Share the joy of the festival together; In the Malay Eid, Other ethnic groups will also send their blessings, Celebrating it all together, Enhance national identity and belonging in the "Selamat Hari Raya" (Happy Eid),

To create a social atmosphere of inclusiveness and unity, Make Singapore a warm home for different cultures to live together and thrive, Set a model for the worlds multicultural harmonious coexistence.

In international exchanges, diplomatic communication and non-governmental exchanges are important ways to realize value exchange. Like two parallel tracks, they are making steady progress in international cooperation. At summits of national leaders and forums of international organizations, they had in-depth dialogues on global issues such as climate change, counter-terrorism and global trade. Representatives of all countries expressed their views and sought convergence of interests and common values, so as to reach consensus, formulate coordinated policies and work together to meet challenges.

In today's changing and increasingly interconnected world pattern, the economies of all countries are like a closely intertwined network, with increasing interdependence, which affects the whole body. Take the international financial field for example. Once a significant economy has economic fluctuations, the impact is like a boulder thrown into the calm lake, and the ripples will spread around the world at an alarming rate. At such a critical moment, the critical significance of national leaders summits and various international organizational forums has become increasingly prominent, and they have become the core platform for countries to work together to overcome economic problems.

The G 20 Summit focused on the formidable challenge of global economic recovery. Representatives from all countries gathered together to express their views and conduct in-depth and detailed discussions on the coordination of macroeconomic policies. In order to inject vitality into the weak economy, some countries cut interest rates to encourage investment and stimulate consumption by reducing the financing cost of enterprises; some countries resolutely implement quantitative easing policy to increase the money supply, just like injecting the dry market to restore liquidity, stabilize the exchange rate, to avoid the haze of global economic recession, and ensure the smooth and orderly economic development of all countries. Among the international organizations, the WTO, just like the mainstay of the international economic order, plays an irreplaceable key role in coordinating the trade policies of various countries and resolving trade disputes. It has carefully formulated unified trade rules, built a solid framework for global trade, and ensured that the goods and services can circulate freely in a fair and just environment, so that countries can give full play to their own advantages, achieve mutual benefit and win-win results, and effectively promote the prosperity of the global economy.

From the perspective of the people, international volunteer services, wonderful cultural exchange years and far-reaching academic visits are flourishing in full swing, just like spring rains nourishing our hearts and quietly nourishing the flowers of international friendship. People will take this opportunity to experience the unique charm of foreign cultures, deeply explore the value support behind them, gradually eliminate the stereotypes in their minds, and sow the seeds of hope for friendly exchanges. Take China's International Chinese Language Promotion project as an example, this project looks like a solid bridge across the cultural gap between China and foreign countries. Millions of learners around the world are immersed in the process of writing Chinese characters, they start a cultural journey through time and space, understanding the depth of the traditional Chinese culture, touch the wisdom of the ancient Chinese civilization; in the daily communication with the Chinese people, truly appreciate the Chinese "friends from afar, happy" hospitality, step by step to understand Chinese culture and values, and establish a deep friendship with the Chinese people. Looking at the Year of International Cultural Exchange, art groups of various countries frequently exchange visits and perform, and colorful art forms such as music, dance and drama are like clever cultural ambassadors, lightly across national boundaries. French Louvre Palace world-famous works in China, those paintings and sculptures bearing the long history of the West and outstanding aesthetics, let the domestic people like in the art hall, close to feel the beauty of western art, immersed in the rich artistic atmosphere; the Chinese Peking Opera troupe visited Europe and the United States, the actors in gorgeous costumes and ink, show the charm of Oriental culture, a skill and a smile with amorous feelings, attracted foreign audience applause, amazed, for the charm of Oriental art.

If we explore the connotation of value exchange, we will find that empathy and empathy are like the light of the soul, guiding human beings to cross the gap of cultural differences. When facing different points of collision, if can try to stand in each other's position, to understand the motivation and cultural background, with an open mind to accept multiple value, human can hand in hand, calmly cope with global challenges, let the world into a warm and harmonious family, like a warm harbor, inclusive accept people of different color, different beliefs.

Take a look at the cross-cultural team collaboration scene in multinational companies. The team members come from all over the world, different countries give them different working habits and values, and the cultural differences are like the gaps that need to be crossed. Fortunately, the company organizes regular cultural exchange activities, and let everyone sit together and open their hearts to share their workplace etiquette and communication styles in their respective countries. In some countries, the workplace atmosphere is relaxed and casual. Employees are encouraged to express their ideas at any time, and others are relatively rigorous and formal, with clear hierarchy and efficient and orderly decisions. When members understand these differences, they can effectively enhance mutual understanding, skillfully avoid conflicts caused by cultural misunderstanding, cooperate with tacit understanding in work, greatly improve work efficiency, and make concerted efforts to contribute to the development of the company. We will also focus on global public welfare projects. Volunteers, regardless of nationality and race, will gather together for the lofty goals of helping the people in the disaster areas and protecting endangered species. When the natural disasters, leaving the devastated ruins, volunteers of different colors stepped up and fought side by side, sweating to carry relief supplies, racing against time to build temporary shelter, in the action deeply experience the respect of different cultures for life, nature, gathered together The power of love, let the light of hope through the haze of disaster, re-illuminate the disaster.

6.3.5 Value Mutual Trust

Trust, just like an invisible but tough link, closely connected to the nodes of human society, and value mutual trust in a deeper focus on trust building based on the common value pursuit, it is like social building indestructible reinforced concrete, build a solid foundation for human cooperation coexistence, to ensure that the enterprise steady progress, let the human society tall.

In the business world, value trust is undoubtedly the key to the long-term development of an enterprise and the prosperity of the industry, which is like the ballast stone on the voyage of an enterprise. Once a bridge of trust based on common values such as quality commitment and social responsibility is established, a business legend can be forged. Apple, for example, has been driven by innovation and extreme user experience. From product design, each line has been repeatedly polished to present simple and beautiful visual enjoyment; to software system development, every operation step has been optimized for countless times to ensure convenient and smooth interactive experience, and every detail is exquisite and elaborate. With this dedication to excellent quality, Apple has won the trust of consumers around the world. Even though the product price is quite high, it is still popular. In the new product release season, fans are willing to give generously for its excellent quality, and the long queue has become a unique scenery on the street. On the contrary, if the lack of value trust between enterprises, cooperation is easy to fall into the mire of suspicion and default, and nightmares such as supply chain fracture and premature cooperation projects will come one after another. Suppliers always worry about overdue payment and blocked capital turnover; manufacturers are concerned about the uneven quality of raw materials and affecting the finished products character. The two sides beware of each other, the road of cooperation is fraught with thorns, difficult to maintain for a long time. In view of this, it is urgent for industry associations and enterprises to strengthen the construction of integrity, and consolidate the foundation of value and mutual trust through effective quality certification, transparent and open information disclosure and other means. Industry associations set strict quality standards, like the "sword of Damocles" on the head of the enterprise, always urge enterprises to meet the standards, and publicly expose their reputation in the industry. The enterprise actively disclose the key information of product composition, production technology, environmental protection measures, so that consumers can buy safely, be honest with partners, and build a solid trust relationship. For example, some cosmetics enterprises publish the ingredient list in detail and carefully mark the efficacy and content of each ingredient. Food enterprises display the real-time monitoring of production workshops through network live broadcast and other forms, so that consumers can witness the whole process of food production, so as to enhance public trust and build a solid root for the sustainable development of enterprises.

By switching to international relations, the value and mutual trust between countries is closely related to world peace and stability, which is like a key safety valve for maintaining the international order. Allies are closely linked based on core values such as common security, democracy and freedom, forming an indestructible alliance of trust and working together to respond to external threats. Looking back at the Cold War period, NATO provided collective security guarantee for European and American countries, and in the tense and treacherous international situation, enabled member states to secure their development and play an important role in regional stability; in the new era, with deep recognition of equal development and the value of South-South cooperation, continuously strengthened economic and trade exchanges, BRICS cooperation is booming, promoted the wave of global multipolar development, significantly enhance the voice of developing countries on the international stage, and played the strong voice of emerging forces. However, it has to be warned that once strategic misjudgment and conflict of values between major countries occur, they can easily lead to crises such as arms race and trade friction, dealing a heavy blow to global welfare. During the Cold War between the United States and the Soviet Union, the two sides fell into the ideological sharp confrontation, invested huge military spending to expand arms, missiles, nuclear war clouds over the world, the people live under the haze of fear; look at the moment, trade protectionism with the rise of slag, some countries try to provoke trade disputes under the guise of values and wave the big stick of tariffs, seriously disrupting the stability of the global industrial chain and supply chain. As a result, many enterprises face difficulties such as loss of orders and shortage of raw materials, and the road to global economic recovery is bumpy. In order to resolve the crisis, countries urgently need to enhance understanding and recognition of values through strategic dialogue, people-to-people exchanges and other diversified channels, cultivate the soil of mutual trust, and make every effort to maintain international peace and stability, keep the world away from the smoke of war and share the fruits of peace and prosperity. As the world's two largest economies, China and the United States in economy, trade, climate, epidemic and other key areas, effectively reduce the risk of misjudgment, honestly negotiate solutions on the trade balance, jointly share the vaccine research and development achievements during the epidemic, actively carry out educational exchange projects, let students go to foreign countries, deeply understand the foreign cultural values, draw knowledge and nutrients in the collision and integration of different cultures, grow into a bridge to promote international exchanges, and reserve strong talents for future international cooperation.

Building value mutual trust is a long-term and systematic project, which urgently needs to make all-round coordinated efforts from individuals, organizations to the country. For the individual purposes, When based on good faith, Practice virtue in your daily life, Keep the promise, and return the money, Seemingly insignificant acts of kindness, In fact, like the stars coming together, To contribute to the social integrity atmosphere; At the social organization level, Public welfare organizations shoulder the responsibility of supervising corporate philanthropy, Track in depth where the money goes, Make sure that every piece of love is accurate, Let the love not be let down, Industry alliances are committed to regulating the internal competition order, Carefully develop the fair competition rules, Prevent unfair competition from destroying the foundation of trust, Maintain the ecological health and order of the industry; In the national dimension, Need to constantly improve the legal and regulatory system, Fraud and breach of contract, Establish a blacklist system for dishonesty, Make it difficult for those to walk, At the same time, to strengthen international cooperation, Police forces work together to pursue transnational crimes across borders, Network security experts unite to fight online fraud, Cracking down on illegal acts that undermine trust, To provide a solid guarantee for building a global network of value and mutual trust, Leading the human society to stride towards a better future on the foundation of mutual trust.

6.3.6 Value Consensus

Value consensus is not an illusory castle in the air, nor can it be a fleeting and overnight achievement. It is like a magnificent spiritual building, which is urgently needed for all countries, nationalities and all sectors of society to be carefully built with full patience and tenacity. Its significance, just like the high sky, yi yi is unripe brightness Polaris, in the surging tide of globalization, digital age torrent rolling forward at present, for human society, the ship accurately illuminate the long voyage, to ensure that humans in the face of many challenges and fleeting precious opportunities, to make steady progress along the correct channel.

From the individual level, everyone is undoubtedly the practitioner of the firm value consensus and the disseminator of enthusiasm. In those seemingly ordinary, trivial moments in daily life, our word and action, every move, are like a brush, or for the value of this grand picture scroll add gorgeous color, or leave indelible stains. A sincere and warm smile, just like the warm spring sun, instantly dispel the heart haze; a goodwill from the heart, passing the compassion and care for others; a soft warm greetings, full of strong human touch. These unremarkable small actions actually contain the universal values of respect, friendship and love, like a babbling trickle, accumulated over time, which can nourish the vast spiritual soil of the whole society and make it full of vitality. On the contrary, negative acts such as indifference, fraud and selfishness are like a cancer hidden in the dark, quietly but extremely destructive eroding the trust foundation that society depends, and become numerous obstacles on the road of value consensus cohesion. Might as well take a look at the warm-hearted scene in the community: warm-hearted residents actively take care of the lonely elderly, tirelessly to help purchase daily necessities, in the elderly body is not timely. Accompany them through the various departments of the hospital. Such kindness, let the old people really feel by the world is still warm, just like a stone into the calm lake, causing layers of ripples, attracted people around to follow, over time, the good fashion of mutual help will become a trend in the community; however, look to the network world, individual netizens are wanton, without hesitation to make false remarks, like the "keyboard man", malicious attacks on others, a time the public opinion into chaos, harmonious communication atmosphere is fragmented, make peoples trust in the network space is greatly reduced, heart fear. Therefore, as individuals, we must always keep self-examination, with a positive and sunny attitude, take the initiative to embrace and practice those values that are beneficial to the harmonious development of the society, so that we can become the value consensus shining and guiding the direction of the boundless dark night.

Focusing on the field of education, schools, families and society should form a close synergy and work together to cultivate a new generation with a sense of value consensus. As a sacred hall of knowledge inheritance and a key position of value shaping, curriculum planning should not only be limited to the one-way teaching of academic knowledge, but also be forward-looking and skillfully integrate far-reaching value education content such as humanistic spirit, social responsibility and global vision. In the history class, the teacher leads the students through time and space, to understand the profound connotation and the great power of pluralistic symbiosis. In the moral and rule of law class, the teacher skillfully guides the students to explore the potential value conflicts and difficult choices behind the social hot issues, and cultivates their sharp vision between right and wrong. Family, no doubt is the first class of children's values sprout to take root, parents words and deeds just like the spring breeze rain, moisten things silent. Parents daily behavior of respecting the old and loving the young, like the seeds of kindness and integrity, quietly rooted in the young hearts of children; the ideal of life and social responsibility is more like a beacon to illuminate the direction of children forward and help. They set up the right life orientation. Expand to the social level, museums, libraries, science and technology museum and so on the public cultural institutions, should fully display its unique education function, carefully held various themes, rich and colorful exhibitions, lectures, and science activities, in unique, fascinating form to the public scientific spirit, humanistic feelings the value concept of yi yi is unripe brightness, if pushed a fan of the door of a new world, broaden the horizons, comprehensive improve the people.

When it comes to enterprises, the pursuit of economic benefits and the pursuit of social responsibilities are not mutual conflict, fire and water, but actually complement each other. The two come together into a strong force to help the stable construction of value consensus. On the one hand, with the two sharp edges of innovative products and optimized services, enterprises break through the difficulties to create solid value for consumers and promote the industry to stride forward and achieve steady economic growth. This process itself is to practice the highly respected value concepts such as enterprising and efficiency. Such as technology enterprises at massive research and development funds, overcome technical difficulties, ingenuity to improve the peoples quality of life, intelligent products, improve life convenience and work efficiency; manufacturing enterprises control the product quality with strict standards, with high quality products in the market reputation, for the benign development of the industry. On the other hand, enterprises actively participate in public welfare undertakings, focusing on many hot social issues, such as environmental protection, education and poverty alleviation, and assistance for vulnerable groups, and give back to the society with practical actions, showing great love and responsibility. Look, some socially responsible enterprises have donated generously in poor areas to build Hope Primary School, for those in difficulties and eager for knowledge to change their fate. Children light up hope and pave the way for growth, while others promote green production to reduce pollution emissions and contribute to the global challenge of climate change. Such economic and responsibility development model, not only can significantly enhance enterprise brand image, make it stand out in the market, won the widespread praise from all walks of life, more can in the enterprise operating every subtle link the value of positive signal, drive the industrial chain upstream and downstream enterprises heart think toward one place, interest makes toward one place, together to participate in the process of value creation and sharing.

Social organizations play a unique and irreplaceable key role in this magnificent journey of value consensus construction. Relying on their inherent flexibility and professional skills, NGOs focus on social pain points and carry out targeted public welfare projects in an orderly manner. Look at the environmental groups, Resolutely launched a river protection action, Carefully organize volunteers to go to each river, They worked pains to clean up the river trash, Like the guardian of the home; Use professional equipment to monitor water quality changes, Check the pulse of the river for its health, With this series of solid actions to arouse the public awareness of environmental protection, Vigorously advocate the harmonious coexistence between man and nature, a value concept containing profound wisdom; Human rights groups have always stood their ground, In order to protect the rights and interests of vulnerable groups, To speak out for the discriminated group, Promoting the value of fairness and justice with the heart is like the seeds sown, Take root and thrive. Industry associations also play an indispensable role in guiding norms, carefully formulate industry self-discipline standards, build a strong line of defense, strictly prevent the breeding and spread of unfair competition, to safeguard the overall reputation of the industry, escort the rights and interests of consumers, and lead member enterprises to move steadily in the value track of integrity and cooperation. On the road, to achieve steady development. All kinds of social organizations are like energetic ties. Through colorful cultural activities, intense academic seminars, and heart-warming volunteer services and other diversified activities, people from all over the world and different backgrounds are gathered together to create a platform for communication and interaction for their ingenuity. In this vibrant world, ideas collide with each other, emotions blend and converge into a warm current, and people can continuously deepen their understanding and identification of each other's values, continuously broaden the originally limited boundary of value consensus, and extend it to a broader world.

Finally, we will expand our vision to a broader international stage. International organizations, governments and civil society organizations must work together to overcome the regional, cultural and political system obstacles to promote the successful conclusion of the global value consensus. Among international organizations, the United Nations is a mainstay. It has taken a strategic position in many key areas, such as maintaining world peace, promoting sustainable development and safeguarding human rights. It has formulated international guidelines and programs of action, and carefully outlined the basic framework for the global consensus on values, like drawing a navigation chart to guide all countries forward. In accordance with their national conditions, governments actively respond to the international call when formulating domestic policies and carrying out international cooperation, assess the situation and skillfully integrate global values into the national development strategy, so that they can take root. At the same time, governments of all countries have made use of diplomatic channels as a communication bridge to strengthen exchanges and cooperation with other countries, seek common ground while shelving differences and reach consensus on many thorny global issues such as climate change, terrorism and the global epidemic, so as to pool the wisdom of all parties and form a coordinated action plan and work together to overcome difficulties. Civil people groups with civil diplomacy unique affinity and appeal, quietly close the distance between the people. International sister-city exchange activities are being in full swing, people in different cities across borders, share each other life, feel exotic customs; cross-border folk art exchange activities, music, dance, painting and other art forms become a bridge of spiritual communication, let the people of different countries in experience to capture the common emotion and value pursuit, as to dispel the fog, eliminate misunderstanding and prejudice, for international cooperation, make the road more smoothly.

The construction of value consensus is undoubtedly a complex and multi-level system engineering, covering individuals, education, enterprises, social organizations and the international community. Only each link close bite, as precision mechanical operation; each participant to burden, with rock firm faith and down-to-earth action, to cultivate, physically practice, enthusiasm to spread the values of human wisdom and wishes, human can build a foundation in the changing, changing world, calmly resolve the crisis, accurately grasp development opportunities, towards a brighter, harmonious and prosperous future steady, to write the magnificent epic belongs to all mankind. It is the eternal theme of the development of human society. It is also the direction that we pass down from generation to generation. It carries the ardent hope and bright dreams from generation to generation, and inspires us to never stop and forge ahead on the long road of pursuing value.

In short, the construction of value consensus value order, mutual learning, value interconnection, value, value trust and value consensus, and many other key elements, they are closely related to each other, together, like a set of close occlusion, efficient operation of gear group, promote the human society in the tide of globalization and the torrent of digital age. Only with the concerted efforts of all countries and nations in the world to build a solid foundation of consensus on values at all levels can mankind calmly resolve numerous crises, accurately grasp development opportunities, forge ahead toward a brighter, harmonious and prosperous future, and write a new glorious chapter of human civilization.

Chapter 7
The Value Revolution of Human Civilization Brought about by the Digital Age

In today's turbulent, rapidly changing world, which is akin to a vast ocean with tumultuous waves, digital technology emerges as a surging, irresistible force, possessing an earth-shaking power that is set to transform every corner of the globe. It acts as an indefatigable, courageous pioneer, penetrating every nook and cranny of human society. Whether it's the bustling city where neon lights cast their colorful glow, reflecting off the glass facades of skyscrapers, or the noisy streets teeming with life, where people from all walks of life hustle and bustle, or the tranquil countryside where the gentle breeze carries the scent of wheat swaying between fields, digital technology leaves its distinct mark everywhere. It's in the smartphones that farmers use to monitor their crops, the tablets that students use in classrooms, and the smart devices that manage our homes.

This technological revolution carries epoch-making significance, much like a bright meteor piercing through the dark night sky, heralding an unprecedented value revolution in human civilization. It has opened a grand curtain to reveal a new era. This revolution, akin to a chain reaction, began with the most basic digital technologies. The latest scientific and technological achievements, like heavy stones, have been thrown into the calm lake of technology, creating sudden waves of innovation that make the entire technological landscape magnificent and send out ripples of splendor across every industry.

Digital technology then strides towards the virtual world, boldly innovating and seamlessly integrating with the real world. It is as if a magical door to a new dimension of time and space has been unlocked, revealing a brand-new world of fantasy and the unknown for humanity. The global economic landscape is being manipulated by a pair of magical, invisible hands, reshaping the universe and transforming the future survival modes of humans. It is akin to breaking out of a cocoon and emerging as a butterfly, a metamorphosis that signifies a profound change.

The majestic power of numbers is redefining our lives in ways we could never have imagined. Algorithms are now the architects of our daily decisions, from the routes we take to work, guided by GPS, to the products we purchase online, recommended by

L. Yan, *Value-Based Civilization*, Understanding China,
https://doi.org/10.1007/978-981-96-9490-7_7

sophisticated recommendation engines. Big data is the new oil, fueling the engines of progress and powering the artificial intelligence that is becoming increasingly adept at understanding and predicting our needs.

In healthcare, digital technology is revolutionizing the way we diagnose and treat illnesses. Wearable devices monitor our vital signs in real-time, while telemedicine allows patients to consult with doctors from the comfort of their homes. In education, online platforms provide personalized learning experiences, breaking down geographical barriers and making knowledge accessible to all. In finance, blockchain technology is creating secure, transparent, and efficient systems for transactions and record-keeping.

The digital transformation is not just about convenience; it's about empowerment. It's about giving voice to the voiceless through social media platforms and enabling small businesses to compete on a global scale through e-commerce. It's about connecting people across continents in real-time, fostering global collaboration and understanding.

However, this digital revolution also brings challenges. It demands that we adapt quickly to new technologies, re-skill, and up-skill to stay relevant in the job market. It raises questions about privacy, security, and the ethical use of data. As we navigate this new digital frontier, we must be mindful of the potential for digital divides and strive to ensure that the benefits of this technological renaissance are accessible to all, regardless of socioeconomic status or geographic location.

In conclusion, the power of digital technology is undeniable and its impact profound. As we continue to harness this force, we must do so with a sense of responsibility and foresight, ensuring that we shape a future that is inclusive, equitable, and sustainable for generations to come.

7.1 Digital Technology Revolution and Value Revolution

7.1.1 Digital Technology Revolution

Digital technology revolution, is undoubtedly a new era can open the magic key, its core essence is computer technology, network communication technology, big data, artificial intelligence, a series of cutting-edge technology in cluster growth and iterative upgrade, they interweave, together, played the passionate movement of science and technology progress. Looking back on here, Computer technology goes along the way, Going from occupying a huge space and having limited computing power, Being as clumsy as a giant, Gradually evolved into a small and portable, excellent performance of the microcomputer, Become a handy tool in people's hands; From the initial single-core processor hard support operation, As hard as a bicycle is walking alone, To now multi-nuclear and even nuclear forces, Like a thousand troops and horses galloping, Easy to cope with complex tasks; From performing only simple basic operations, Transformation to an advanced stage of simulating complex realistic scenarios and solving difficult problems, The computing speed is even more exponentially soaring, Like a rocket flying as fast, It lays a rock-solid foundation for the efficient processing of massive data. The development process of network communication technology is also magnificent, remote thinking in the early days, people through the dial-up Internet, endure like the speed of a snail crawling, web page loading slowly, information Transmission lag, every waiting as if suffering, and now, we have striding into 5G and look forward to the future 6G era, information can quickly between every corner of the world, completely broke the gap between regional if cutting, let the global village really become a reality, no matter where in the world, can be connected in real time. Big data technology just like a unique eye "intelligent strategist", it will be fragmented, massive and messy information clever integration, deep mining, as in the vast data, accurately extract hidden rules and trends, for the enterprises strategic decision-making, social efficient governance to provide forward-looking guidance, become the wisdom of the brain. Artificial intelligence is given the machine to simulate the magic of human intelligence, in the medical field, from image recognition technology power accurate diagnosis of disease, quickly identify the subtle lesions of medical imaging, as to the doctor installed a pair of "perspective eyes", to the natural language processing achieve smooth dialogue between man and machine, auxiliary doctors understand patients, like close medical assistant, expanding the boundary of human ability, for medical career development.

These landmark technological breakthroughs have set off stormy waves in all fields of human society and brought about earth-shaking changes. In the medical industry, remote medical with the aid of hd video transmission technology and real-time data monitoring system, set up a remote bridge between patients and top medical experts, the bridge across the mountains, let those in traffic inconvenience, lack of medical resources of patients, never leave home can enjoy authoritative expert consultation services. As during the COVID-19 pandemic, Online consultation platforms have mushroomed, It provides an indispensable and key support for epidemic prevention and control and reasonable distribution of patients, Effectively alleviating the shortage of medical resources, Become a strong defense line to protect people's health in a special period; Smart medical devices are even more brilliant, Using various precision sensors to continuously collect human body data, Combined with advanced AI algorithms for in-depth analysis, Early warning of disease risk, Like the popular wearable bracelets on the market, Real-time monitoring of physiological indicators such as heart rate and blood pressure, Like a caring personal health guard, Always guarding people's health, For People's Daily life escort. In the field of manufacturing, the intelligent factory under the wave of Industry 4.0 has become a new benchmark for the development of the industry, with robots with high precision mechanical arm and Accurate program setting, in the complex process can complete the parts assembly task, automatic production line to achieve seamless connection, material distribution, processing and manufacturing, finished product testing and other links at one go, the production efficiency is geometric multiple surge, the production quality is more stable and reliable. Take BMW automobile factory as an example, its intelligent robot can flexibly adjust the production process according to different order needs, quickly switch the production mode, realize customized production accurately and efficiently, and meet the increasingly diversified needs of consumers. It seems to have the magical "deformation" ability to adapt to the rapid changes of the market. The field of education has also benefited greatly, The online course platform is like a treasure house of global quality education resources, Students with just a click of the mouse, Can cross the boundaries of time and space, Listen to the wonderful lectures from famous masters from all over the world, Whether with cutting-edge scientific knowledge, Or the ancient history and culture, You can catch it all all together, Knowledge dissemination has since broken through the traditional time and space limitations, Opened a new era of universal learning; Interactive learning software is even more creative, Through virtual experiments, students are able to operate experimental equipment, observe experimental phenomena, Intelligent q & A system is at any time for students to dispel doubts, Deeply stimulated the students desire to explore the unknown, Such as the moOCs network. It has attracted tens of millions of students to take different professional courses, providing a broad platform for them to broaden their knowledge horizons and pursue their academic dreams, and becoming an effective partner for students on the way to seek knowledge.

The impact of the digital technology revolution is far more than just changing the way of production and life. It is also like a quiet but highly skilled craftsman, quietly reshaping the human thinking pattern. In this wave of rapid iteration and update, people gradually adapt to the rapidly changing rhythm, and learn to actively embrace uncertainty, and innovation and change has become the passionate theme of The Times. Enterprises have abandoned the traditional linear research and development mode, and have adopted agile development strategy, closely track market dynamics, quickly respond to market demand, and launched products that meet the needs of consumers. Under the impact of information flood, a person gradually practices a pair of eyes, learns to screen effective information, and uses various digital tools. The concept of lifelong learning is deeply rooted in the hearts of the people, and becomes a powerful weapon for people to forge ahead in the tide of The Times, as if they have the ability of continuous evolution. However, we must also clearly realize that digital technology revolution is like a double-edged sword, bring many welfare at the same time, also caused a series of difficult problems, such as privacy problems, personal information in cyberspace if naked, bring many problems to people's lives, as privacy "magino line" is light Easy to break through; the Internet addiction is increasingly serious, some people addicted to the virtual world, physical and mental health erosion, as if in the invisible "swamp"; digital divide is widening, between urban and rural areas, between different age groups, between developed and developing countries in digital technology access and application ability, like an insurmountable "digital barrier". These problems to cause the whole society attention, through the construction of perfect ethics system, strengthen industry self-discipline, as built solid "moral wall", while perfecting the laws and regulations and policies, strengthen regulatory enforcement, guide the digital technology to good development, to ensure that the magnificent revolution can benefit all human, rather than become a minority "patent", let the light of the digital equal shine every Angle.

7.1.2 Artificial Intelligence and Human Civilization: Evolution and Reconstruction in the Technological Wave

In the year 2025, the world's attention was captivated by the Paris AI Summit, an event that would go down in history for its groundbreaking demonstration by Neuralink. During this remarkable showcase, a human volunteer, with a small, unobtrusive device implanted in their brain, was able to seamlessly access a vast repository of AI-generated medical knowledge. This device, a sophisticated brain-computer interface, allowed the volunteer to query complex medical questions and receive instantaneous, accurate responses directly into their mind, as if the AI's knowledge was their own.

This moment was more than just a technological milestone; it symbolized the dawn of a new epoch where silicon-based intelligence intersects with human cognition. It was a profound moment that began to reshape the very essence of what it means to be human. The implications were vast and far-reaching, suggesting a future where the boundaries between human and machine intelligence could become increasingly blurred.

As global investment in AI technology continued to surge, surpassing an annual figure of $2.3 trillion, the pace of innovation was staggering. The capabilities of machines had evolved to the point where they outperformed humans in a multitude of domains, ranging from the intricate task of protein folding to the strategic complexities of gaming. In these fields, AI had not only caught up with human expertise but had surpassed it, achieving feats that were once considered the exclusive domain of human intellect.

The acceleration of this intelligent revolution raised profound questions about the trajectory of human civilization. What would it mean for society when AI could diagnose diseases with greater accuracy than the most skilled physicians? How would the job market adapt when machines could strategize and outmaneuver the best human minds in gaming and beyond? These questions sparked intense debate among ethicists, technologists, and policymakers, as the world grappled with the potential benefits and risks of this rapidly advancing technology.

The demonstration at the Paris AI Summit was a pivotal point that highlighted the potential for a symbiotic relationship between humans and AI. It suggested a future where the limitations of the human brain could be augmented by the computational power of machines, opening up new horizons in education, healthcare, and problem-solving. Yet, it also raised critical questions about privacy, autonomy, and the ethical use of such powerful technology. As humanity stood on the brink of this new era, the world watched with a mix of awe and apprehension, eager to see how this intelligent revolution would unfold and what it would ultimately mean for the future of humankind.

The roots of artificial intelligence trace back to the 1956 Dartmouth Conference, a seminal event in the history of technology where a group of visionary scientists and mathematicians gathered to explore the possibilities of creating machines that could simulate human intelligence. It was during this conference that John McCarthy, a brilliant computer scientist, first coined the term "artificial intelligence" and proposed the symbolic manipulation of logic as the path to achieving machine intelligence. This concept laid the groundwork for the field of AI, setting the stage for future developments and research.

Early systems like the General Problem Solver (GPS), inspired by McCarthy's ideas, attempted to mimic human reasoning through formal rules. These systems were designed to solve a wide range of problems by breaking them down into smaller, more manageable parts and applying logical rules to find solutions. However, their limitations soon became apparent as researchers realized that the complexity of human thought could not be fully captured by simple rule-based systems. The inherent difficulties in replicating the nuances of human cognition highlighted the challenges of creating truly intelligent machines using purely symbolic methods.

The 1974 Lighthill Report marked a significant turning point in the history of AI. Authored by James Lighthill, a prominent British applied mathematician, the report critically assessed the progress of AI research and concluded that it had fallen far short of its ambitious goals. This scathing critique led to a decade-long funding drought, often referred to as the "AI Winter," during which government and private funding for AI research was severely curtailed. This period of stagnation and reduced investment underscored the difficulties faced by the field and the need for new approaches to achieve the elusive goal of machine intelligence.

The 1980s witnessed a resurgence in AI research with the rise of connectionism, a paradigm inspired by the structure and function of the human brain. Innovations such as backpropagation algorithms and parallel distributed processing models revolutionized the field by enabling systems to learn from data rather than relying solely on hand-coded rules. These advancements allowed AI systems to improve their performance over time by adjusting their internal parameters based on feedback from their environment.

Japan's Fifth-Generation Computer Project, launched in the early 1980s, exemplified this new wave of optimism in AI. The project aimed to create a new generation of computers capable of natural language processing, problem-solving, and expert-level reasoning. The ambitious goals of the project included the development of supercomputers that could emulate human cognitive processes, leading to significant advancements in computing power and software capabilities. However, despite some notable successes, the computational resources of the era proved insufficient for the large-scale implementation of neural networks, leading to another wave of disillusionment by the mid-1990s.

The turning point for modern AI came in 2012 with the victory of AlexNet, a deep convolutional neural network, in the ImageNet competition. This competition, which involved a large-scale visual recognition challenge, had long been a benchmark for the field of computer vision. AlexNet's impressive performance demonstrated the power of deep learning, a subset of machine learning that utilizes neural networks with many layers. Convolutional neural networks (CNNs), a type of deep neural network particularly well-suited for image recognition tasks, revolutionized computer vision by achieving accuracy levels previously deemed unattainable.

This breakthrough unleashed a torrent of advancements in AI. In 2016, AlphaGo, an AI program developed by Google DeepMind, made headlines around the world by defeating Lee Sedol, one of the greatest Go champions of all time. AlphaGo's victory showcased the strategic prowess of reinforcement learning, a technique that allows AI systems to learn optimal actions through trial and error. By playing against itself and learning from millions of games, AlphaGo developed strategies that were both creative and highly effective, surpassing human expertise in the complex game of Go.

In the same year, the launch of GPT-3 further demonstrated the enormous potential of AI in the field of natural language processing. GPT-3 (Generative Pre-trained Transformer 3) is an advanced language model capable of generating coherent, fluent, and creative text, often indistinguishable from the work of human authors in many cases. The advent of GPT-3 not only propelled the development of natural language processing technology but also sparked widespread discussions about the extensive applications of AI in literary creation, content generation, and dialogue systems, among other areas. These breakthrough advancements indicate that AI technology has reached a new height, beginning to exhibit its powerful application potential across multiple domains.

Today, the field stands at the threshold of general artificial intelligence (AGI). Multimodal models like Google Gemini integrate text, image, and speech processing, achieving human-level performance in diverse tasks. Quantum computing further accelerates progress, with D-Wave's 3000Q system solving combinatorial problems at speeds 10^{12} times faster than classical machines. These developments have sparked debates about the "technological singularity," a hypothetical point where AI surpasses human intelligence and drives irreversible societal change.

The impact of AI on the global economy is both transformative and disruptive. In manufacturing, robotic systems now account for 322 units per 10,000 workers globally, automating 90% of tasks in automotive assembly lines. Germany's Industrie 4.0 initiative, for instance, uses AI to optimize supply chains, reducing inventory costs by 40%. Meanwhile, AI-driven innovation has birthed entirely new industries—companies like OpenAI and DeepMind now command valuations exceeding $100 billion, reshaping market dynamics.

However, this progress comes with societal costs. The World Economic Forum estimates that 41% of jobs are at risk of automation by 2025, particularly in routine and cognitive roles. Japan's "Robot New Strategy" plans to replace 3.5 million care-giving positions with AI systems, raising concerns about workforce displacement. To mitigate these effects, governments are exploring universal basic income (UBI) models, with pilot programs in Canada and Finland showing promising results in reducing economic inequality.

Despite these challenges, AI remains a critical engine of growth. McKinsey predicts that AI could contribute $15.7 trillion to the global economy by 2030, primarily through increased productivity and new service offerings. Central banks are already integrating AI into economic forecasting, improving the accuracy of monetary policy decisions by 30%. The rise of algorithmic trading platforms, which now account for 60% of stock market transactions, underscores AI's role in reshaping financial systems.

The integration of AI into governance is redefining the relationship between citizens and the state. Singapore's "Smart Nation 2030" initiative exemplifies this trend, using AI to analyze 1 billion daily social media posts to predict public sentiment and optimize policy outcomes. Similarly, China's "City Brain" project in Hangzhou processes real-time data from 800,000 surveillance cameras to reduce traffic congestion by 15%. These systems enhance administrative efficiency but also raise concerns about privacy and algorithmic bias.

Legal frameworks are struggling to keep pace with technological advancements. The EU's Artificial Intelligence Act, the world's first comprehensive AI regulation, classifies systems into four risk tiers, mandating transparency and human oversight for high-risk applications like facial recognition. Meanwhile, the International Court for AI Disputes (ICAID) has adjudicated landmark cases, such as assigning liability in self-driving car accidents, establishing precedents for algorithmic accountability.

Ethical considerations are central to these debates. The 2025 Cambridge Analytica scandal, where AI was used to manipulate electoral processes, highlighted the risks of algorithmic influence. To address this, organizations like the Future of Life Institute promote the "Beneficial AI" movement, advocating for principles of transparency, fairness, and human control.

AI is fundamentally altering cultural production and consumption. Platforms like MidJourney and DALL-E generate artwork indistinguishable from human creations, challenging traditional notions of authorship. The 2024 Colorado State Fair controversy, where an AI-generated piece won a top prize, sparked debates about the role of human intent in art. Similarly, AI writing tools like GPT-4 now contribute to 34% of digital content creation, raising questions about originality and intellectual property.

Education is undergoing a parallel transformation. Finland's "Human-AI Classroom" pilot program uses brain-computer interfaces to augment student learning, achieving a 30% improvement in retention rates. Universities like Stanford are integrating AI tutors into curricula, providing personalized feedback to millions of online learners. However, this shift also risks devaluing human teachers, prompting calls for new models of collaborative instruction.

The media landscape has been irrevocably changed by AI-driven algorithms. Social media platforms now use machine learning to curate content, creating "filter bubbles" that reinforce user biases. A 2025 MIT study found that algorithmic recommendations increase ideological polarization by 47%, threatening democratic discourse. Countering this trend, initiatives like the EU's "Digital Literacy Act" aim to educate citizens on media manipulation techniques.

As AI permeates every aspect of society, humanity faces a defining choice: to embrace symbiotic coexistence or risk becoming obsolete. The concept of "human-AI collaboration" is gaining traction, with systems like IBM's Watson assisting doctors in diagnostic decisions and legal AI tools streamlining contract reviews. However, achieving this balance requires addressing critical challenges.

Technical barriers remain significant. Despite advancements, AI systems often lack explainability, making it difficult to attribute responsibility for errors. The

2025 Boeing 797 crash, caused by an AI flight control system's miscalculation, underscored the dangers of overreliance on opaque algorithms. To mitigate this, researchers are developing "interpretable AI" techniques that provide insights into decision-making processes.

Ethical dilemmas also persist. The development of lethal autonomous weapons (LAWs) has sparked international outrage, with 38 nations signing the 2025 Hague Convention banning their deployment. Meanwhile, debates rage over the ethical implications of AI in healthcare, particularly in end-of-life decision-making. The World Medical Association's 2025 guidelines require human oversight in all AI-assisted medical procedures.

Looking beyond Earth, AI is enabling humanity's expansion into space. SpaceX's Starship, guided by advanced AI navigation systems, has completed multiple Mars missions, establishing the foundation for interplanetary colonization. These ventures rely on AI to manage life support systems, grow food in harsh environments, and maintain infrastructure—tasks previously impossible without human intervention.

However, the colonization of space raises new ethical and existential questions. The 2025 Moon Treaty, which governs the exploitation of extraterrestrial resources, includes provisions requiring AI systems to prioritize human safety. As we venture further into the cosmos, the role of AI will evolve from assistant to co-pilot, shaping the destiny of humanity's diaspora.

The intelligent revolution represents both humanity's greatest opportunity and its gravest threat. As AI systems surpass human capabilities in increasingly complex domains, we must resist the temptation to abdicate responsibility to machines. Instead, we must harness this technology to augment our collective potential, ensuring that progress aligns with ethical values and human dignity.

China's "Digital Community with a Shared Future" initiative offers a blueprint for global cooperation, emphasizing technology as a tool for inclusive development. By investing in education, fostering innovation, and strengthening governance frameworks, we can navigate the challenges of this new era.

Ultimately, the story of AI is not one of machines replacing humans but of humanity redefining itself. This profound transformation is not merely about the evolution of technology; it's about the evolution of our very essence. As we stand on the precipice of AGI, the advent of Artificial General Intelligence, we are at a crossroads where the choices we make today will have far-reaching consequences. These choices will determine whether we harness the power of AI to create a future where technology serves as a tool for enhancing human capabilities, fostering innovation, and creating a world of shared prosperity. Or, conversely, we could find ourselves descending into a technological dystopia, where the very tools we created to improve our lives instead lead to widespread unemployment, social inequality, and a loss of human connection.

The power to shape this destiny lies not in algorithms, no matter how sophisticated they may be. It lies in the timeless human values of empathy, wisdom, and foresight. Empathy is the ability to understand and share the feelings of another, a quality that must guide our interactions with AI. As we develop and deploy AI systems, we must consider the impact on all members of society, ensuring that these technologies do not exacerbate existing inequalities but rather work to bridge the gaps between us.

Wisdom, the ability to make sound judgments and decisions, is crucial in navigating the complex ethical landscape that AI presents. We must approach AI development with a deep understanding of the potential risks and benefits, and with the foresight to anticipate and mitigate negative consequences. This wisdom should inform the policies and regulations that govern AI, ensuring that they are robust enough to adapt to the rapid pace of technological change while being flexible enough to allow for innovation.

Foresight, the capacity to envision future possibilities and outcomes, is essential in preparing for the long-term implications of AI. We must look beyond immediate gains and consider the broader societal impacts, including the potential for job displacement, the need for re-skilling workers, and the ethical treatment of AI entities. By exercising foresight, we can create a roadmap for a future where AI is integrated into society in a way that maximizes benefits and minimizes harm.

In conclusion, the story of AI is a narrative of human potential and responsibility. It is a story that we, as a society, are writing together. This tale is not just about the development of machines that can think and learn, but also about the profound impact these technologies have on our lives, our work, and our understanding of intelligence itself. The chapters ahead will be shaped by our collective actions, guided by our shared values, and ultimately, they will reflect the kind of world we choose to create. As we stand on the brink of this new era, let us commit to a future that is not only technologically advanced but also deeply humane.

As we delve into the intricacies of artificial intelligence, we must recognize that it is not merely a tool for efficiency or a means to an end, but a reflection of our own aspirations and fears. AI has the potential to revolutionize healthcare by providing personalized treatments, to transform education by adapting to individual learning styles, and to enhance our daily lives through intelligent assistants and smart devices. Yet, with this potential comes a great responsibility to ensure that these technologies are developed and deployed ethically and equitably.

The chapters ahead will be written by the policymakers who create regulations to protect privacy and prevent misuse, by the educators who prepare the next generation for an AI-driven economy, and by the innovators who push the boundaries of what machines can achieve. Our shared values, such as fairness, transparency, and accountability, must guide these actions. We must ask ourselves how we can foster an environment where AI is used to uplift rather than to divide, to empower rather than to control.

Ultimately, the story of AI will reflect the kind of world we choose to create. Will we harness the power of AI to solve some of the most pressing challenges of our time, such as climate change, poverty, and disease? Or will we allow it to exacerbate existing inequalities and create new ones? The decisions we make today will determine the legacy we leave for future generations.

As we stand on the brink of this new era, let us commit to a future that is not only technologically advanced but also deeply humane. Let us strive for a world where technology serves the common good, where AI is used to foster understanding and compassion, and where the benefits of innovation are shared by all. By embracing our collective responsibility and working together, we can shape a future that reflects the best of what it means to be human.

7.1.3 Value of the Internet

Value of the Internet, as a new paradigm of digital age value, just like a bright star rising in the digital sky, it is based on block chain decentralized technology architecture, with a disruptive power, completely break the traditional Internet is given priority to with information transmission inherent mode, open up a realization value directly, safe, efficient circulation of new channels, as if on the traditional Internet "old map" found a new "treasure route". Block chain technology, like building a digital world trust cornerstone "magic blocks", its unique distributed books technology, makes the data as scattered storage in many strong "fortress"—node, has the not tamper with, the entire traceability characteristics, just like for each transaction information with a layer of indestructible "protective a", to ensure the authenticity and reliability of transaction information, let the trust of the digital world has a solid support. COINS as the original application of block chain technology, although since its birth was controversial, the price fluctuations, illegal trade use caused many questions, but its success shows in the case of no trust the third party intervention, the feasibility of value transfer and innovation, for the development of the subsequent value of the Internet provides valuable exploration experience, as lit in the dark. A "bright light" of hope. Smart contract is like a loyal, automatic execution of "electronic contract" housekeeper, the terms of the contract in the form of code accurate coding, once the preset conditions meet, can automatically trigger execution, without human intervention, greatly improve the efficiency of the transaction and, reduce the risk of default, as if to trade with intelligent "navigator". In the field of cross-border trade finance, smart contract can be based on the goods to port, qualified key information, instant complete payment process, abandoned the traditional trade cumbersome such as exchange, clearing intermediate link, not only save a lot of time cost, also can effectively avoid trade disputes caused by human factors, as to cross-border trade "wheel" coated with smooth "lubricating oil.

The value Internet took the lead in the financial field, just like a spring breeze, blowing a new chapter of inclusive finance. Decentralized financial (DeFi) platform booming, the platform for the majority of users provides an unprecedented financial services convenient channel, users need not rely on Banks and other traditional financial intermediary, only through encryption wallet, can easily participate in diversified lending, financing, trading financial activities, as if for the public opened a "new door" to financial freedom. In remote areas such as Africa and Southeast Asia, Many farmers lack collateral and lack access to remote financial services, Facing the long-term financing difficulties, Now, with small DeFi loans, They have timely access to funds to buy agricultural materials, Developing agricultural production, Improve the living conditions, Like a long drought, Inject vitality into the local economic development; Cross-border payment field is also due to the major changes in block chain technology, Traditional cross-border payment processes are cumbersome, Involving many banks, clearing houses, High fees and a long time to arrive, The cross-border payment scheme represented by Ripple Company, Using block chain technology to complete cross-border transfers within seconds, Costs only a few tenth of the traditional way, Inject strong impetus into global trade and financial financing, It's like setting up global money Flow of the "high-speed channels.

In terms of content creation and intellectual property protection, the value Internet has endowed creators with new opportunities, just like a light for creators dreams. Not homogeneity the emergence of tokens (NFT), for digital art, music, literature and other works provides a unique identity—approval, artists will create their works after casting into NFT in block chain platform sales, every circulation of work transactions, creators can get the corresponding royalties, which greatly stimulated the creators of creators, let their hard work get real return, as if for the creator on the "double crown" of honor and wealth. The digital paintings of encryption artist Beeple is expensive, this landmark event is like throwing a boulder in the calm lake, triggered the global art circle from the edge of the digital art to the mainstream vision; writers can also put the works on the copyright chain, accurately track the spread of the block chain technology, in the infringement, create a healthy and orderly ecological environment for knowledge creation, encourage more excellent works, as built a solid "protective fence" for the garden of knowledge.

From the perspective of social governance, the value of the Internet helps to enhance the transparency and credibility of social operation, just like injecting a "shot in the arm" into social governance. Government departments use block chain technology to open government data, such as charity funds flow, land transfer records, the public just a click of the mouse, can query supervision at any time, let the public power run in the sun, effectively eliminate power rent-seeking, corruption, such as the government with a transparent "coat"; after the public welfare project chain, donors can know the details of each donations are clear, which greatly stimulated public enthusiasm, let love can accurately pass to the people in need, gather social positive energy, like a trickle into love "ocean".

However, we must also face up to, the value Internet is still in the early stage of development, as young as a newborn baby, facing many growing pains, such as the technical bottleneck has not yet breakthrough, transaction speed, expansion capacity, as if the baby; the regulatory policy is still toddling, some innovative applications in the law, causing market chaos, like in the "sea" navigation; the market volatility, crypto asset prices ups and downs, the protection of investor rights and interests of challenges, as if in the turbulent "financial storm". These problems urgently need for all walks of life to work together to increase investment in research and development to overcome technical problems, the government, industry organizations, enterprises and other parties to improve the regulatory system, guide the rational development of the market, release the huge potential of the value Internet, let it create more benefits for human society, just like carefully protecting seedlings to grow into towering trees.

7.1.4 Value Reconstruction: The Reconstruction of the New Global Value Chain System

The digital age behaves as an unforgiving yet innovative sculptor, relentlessly chiseling away at the traditional global value chain system—a sculpture honed over years of development—propelling the transformation towards a novel value chain paradigm, akin to a devastating artistic upheaval. With each digital wave, the landscape of global commerce is being reshaped, carving out new pathways for efficiency and innovation.

On one side, the digitalization has led the manufacturing sector to embrace a notable shift towards decentralization and localization. As a vanguard of digital manufacturing, 3D printing technology liberates product production from the constraints of traditional centralized mass production, enabling on-demand manufacturing close to the consumer, tailored to individual preferences. This transformation is akin to a seismic shift in the industry, where the power of production is no longer confined to a few centralized hubs but is distributed across a network of smaller, more agile units. This significantly slashes the expenses associated with long-distance transport and storage, bringing the "magic wand" of production into the hands of consumers. Imagine a world where a consumer in New York can design a unique piece of jewelry online and have it printed in a local facility, delivered to their doorstep within days, all while reducing the carbon footprint associated with traditional manufacturing processes.

Consider Adidas, which astutely latches onto technological advancements, establishing 3D printing facilities for sports footwear across the globe. The brand has become a pioneer in leveraging this technology to cater to the dynamic needs of its customers. It adeptly tailors its offerings to local market trends, responding swiftly to consumer demands, catering not only to the pursuit of fashion and personalization but also mitigating supply chain risks with agility, much like "nimble running shoes" that adapt to the terrain. Adidas's Speedfactory concept is a testament to this, where shoes are produced close to the end consumer, ensuring that the latest designs and technologies are quickly available in the market.

The industrial Internet acts as a potent "bond," facilitating the dynamic allocation of production factors on a global scale. It's like a vast, interconnected web that allows for real-time communication and coordination between machines, people, and processes. Advanced film research and development, leveraging the deep technological reserves of Europe and America, along with top-tier scientific talent, secure a leading position in the industry. These regions are hubs of innovation, where breakthroughs in materials science and engineering are constantly pushing the boundaries of what's possible in manufacturing.

Meanwhile, the packaging and testing stages capitalize on cost and talent distribution factors, allowing for a flexible configuration in regions like Asia, maximizing comparative advantages and breaking free from the geographical constraints that once dictated where products could be made most efficiently. In countries like China and Vietnam, labor costs are lower, and there's a skilled workforce ready to take on the meticulous tasks of packaging and quality assurance. This regional specialization allows companies to optimize their supply chains, reducing costs and improving turnaround times.

Furthermore, the digital age has ushered in an era of big data analytics, where vast amounts of information can be processed to predict market trends, consumer behavior, and potential disruptions in the supply chain. This predictive capability is like having a crystal ball that guides businesses through the uncertainties of the market, allowing them to make informed decisions and stay ahead of the competition. The digital age is not just reshaping the physical aspects of manufacturing but is also transforming the very way we think about and manage global commerce.

On the other hand, digital services in the rise of digital services in the background of the digital era, like a dark horse, rapidly become a key force in the value chain. Software outsourcing, cloud computing services, digital content creation and other fields are booming, showing strong vitality and potential, just like a "charge force" of the digital economy. India's software outsourcing industry relies on massive engineer resources, mature and standardized development process, Important place in the global software outsourcing market, From the early basic work of pure "code writing", Gradually advanced to the system architecture design and other high value-added links, Increasing its position in the value chain, Has contributed to India's economic growth, Like setting up a solid "Indian ladder" on the international software stage; China's digital cultural and creative industry has also made remarkable achievements, Output of animation, games, online articles and other works are popular overseas, Domestic games represented by "The Original God", with their exquisite graphics, rich plot and innovative gameplay, So of money around the world, Behind it is the cultural creativity, technology research and development, operation and promotion and other whole chain of close cooperation to achieve the value appreciation, Rewriting the global cultural and trade landscape, Highlighting the soft power of Chinese culture, As if to the world a hand out a beautiful "cultural card".

As the "leading role" on the global economic stage, the strategic transformation of multinational companies accelerates the process of value chain restructuring. Traditional giants are become digital, Actively embrace change, Walmart has invested heavily in building an e-commerce platform, Integrate online and offline resources, Optimize supply chain digital management by using big data analysis, Accurately grasp the consumer demand, Realize inventory optimization and efficient distribution, Enhance the competitiveness of omni-channel retail, As if for the traditional retail installed with the digital "wings"; Technology upstarts like Amazon, Relying on the powerful advantages of cloud computing services and big data analysis, Reverse to cut into the field of physical retail, workshop, logistics and distribution, Data-driven synergies of all links of the value chain, Through intelligent storage management and accurate recommendation algorithm, Control the voice of industrial ecology, Continue to expand the business boundaries, Lead the trend of industry development, As if the "leading flagship" of the digital economy.

In this magnificent wave of reconstruction, the developing countries ushered in the precious opportunity of curve overtaking, just like finding a new shortcut in the corner of the track. However, opportunities and challenges coexist, developing countries are also facing many difficulties, such as weak digital infrastructure construction, insufficient network bandwidth, lack of data center, restricting the development of digital economy, as if the digital talent shortage, shortage of high-end compound talents, difficult to meet the needs of industrial upgrading, like the lack of digital chariot "skilled driver"; serious lack of international rule-making rights, in the passive international digital trade rules game, as if "back layout" in the international chess game. In order to promote the global value chain towards a new journey of balanced and sustainable development and let all countries share digital dividends, international organizations and governments need to work together to establish fair and inclusive digital trade rules, strengthen the assistance for digital infrastructure to developing countries, help them to fill up their shortcomings through technology transfer and financial support, just as the digital vehicle "refueling maintenance" for developing countries; strengthen digital talent training, open relevant courses in college education, vocational training and other fields, cultivate professional talents to meet the needs of the digital age, help development and help developing countries. Moving forward in the restructuring of the global value chain, as if to cultivate the "navigator of its digital journey.

7.1.5 Value Twin: The Meta-Universe of Parallel Worlds

The universe, as the most imaginative and dreamy color in the digital age, is like a amazing concept leaping from the pen of science fiction masters, just like a powerful and magical "dream maker", building a highly virtual parallel space and time that is closely connected with the real world. It is like an indestructible bridge between reality and dream, cleverly triggering the deep twin of the value system, as if gently pushing open an "arbitrary door" leading to the dream wonderland, leading people to leisurely step into a new world full of infinite possibilities and fascinating. It is like a knowledgeable scholar, integrating the essence of virtual reality (VR), augmented reality (AR), artificial intelligence, block chain and other cutting-edge technologies, and using these technologies as a masterstroke and gorgeous paint to carefully create an immersive and interactive virtual scene. Only people easily wear corresponding equipment, can be like moment through time tunnel, suddenly in a realistic to amazing, if immersive virtual world, social, work, entertainment, learning and other activities can be freely in this, open the new dimension of human life, as if have the second life, for peoples life, Change beyond imagination.

In the social field, the meta-universe reshapes the interpersonal communication mode, giving people unprecedented new social experiences. Virtual social networking platforms, such as Decentraland, users have changed into the streets and alleys of the virtual city, holding a grand, extraordinary, lively and noisy virtual parties with great interest. In that dreamlike virtual space, Friends from all over the world, of different colors and different cultural backgrounds laugh and gather together, gaudeamus, Dynamic sound of music, cheerful laughter interwoven and echoed together, Let you feel like you are in a fantastic "digital party island", Enjoy the joy of crossing the regional restrictions, no distance; When people participate in the wonderful and magnificent art exhibitions, Can easily overcome the obstacles of geographical distance and cultural differences, Real-time interactive communication, Share your unique and profound understanding of art works freely, It is like starting an artistic "journey around the world" in the cloud, Let the bright charm of art bloom in the virtual world more gorgeous and dazzling. In the special period when COVID-19 was rampant and offline social interaction was difficult, people's communication was blocked by masks and ruthlessly blocked by physical distance, while Yuanyuan Society. The platform is like a warm dawn breaking through the darkness, Provide unique social comfort, Make it easy to expand the breadth and depth of their social circle without leaving home, Steady maintain the emotional bond, To become a warm harbor for people to comfort their hearts and communicate their emotions in times of hardship; The strong rise of virtual idols is a dazzling highlight in the social field of the universe, Based on AI driven, 3D modeling and elaborate idols, At the holographic concert, The stage effect is shocking and dazzling, Dazzling lighting, extreme special effects, Fans are immersed in it, Chcheer and shout, Their enthusiasm was like a raging fire, Seems like trying to ignite the whole virtual space in an instant. And these virtual idol also with its own unique charm, endorsement of all kinds of brands, attracted huge fans, successfully created a unique emotional ties and rich commercial value, heroic open idol industry era, for a slightly dull boring entertainment industry development into a strong new vitality, let the entertainment industry to glow vitality and infinite live.

In the work scenario, the meta-universe creates a new paradigm of remote collaboration, bringing revolutionary upheaval to the workplace ecology. With the help of virtual offices and remote collaboration tools, employees have become digital bodies and attend transnational meetings calmly. If they are on the scene, they can clearly see colleagues every twinkle and smile and every move, and face to face communication without obstacles, smooth and free. In collaborative project development, you can real-time Shared documents, operate the virtual model, completely broke the space-time boundaries, blurred the boundary between reality and virtual work, employees need not worry about physical commuting, no longer in the crowded traffic spend a lot of precious time and energy. This not only greatly reduces the operating costs of the enterprise, such as saving office space rental expenses and transportation expenses, but also significantly improves work efficiency, allowing the global team to work closely together, as if magically eliminating the regional barriers, so that everyone is in the same room, side by side. Just like Microsofts Mesh for Teams platform, let the global team like living in a room, efficient collaborative office, whether passionate brainstorming, or difficult projects, can be as smooth; digital twin factory clever profit With yuan universe technology, the entity factory accurate real-time mapping to virtual space, engineers through virtual interface can remote monitoring, debugging equipment, if have a pair of "perspective", found the hidden trouble in advance, timely optimization of production process, save day to prevent fault risk, ensure production continuity, greatly improve the quality of production, for manufacturing transformation and upgrading off wings, boost traditional manufacturing strides towards intelligent, efficient new platform.

In the education and training section, Yuanyuan provides a new immersive learning experience and opens a new chapter of knowledge inheritance. In the history class, With the help of sophisticated equipment, Like traveling through a time tunnel, Coming to the scene of the ancient civilization, To see the spectacular spectacle of historical events, To truly feel the vicissitudes of history and the profound deposits, As if he was the witness of the waves of history; In the astronomy course, Students in the vast expanse of the starry sky, Close look of the wonderful trajectory of celestial movements, Explore the deep mysteries of the universe, The dazzling stars are as close at hand, within reach, Greatly stimulate their infinite reverie and desire to explore the universe; Medical students through a virtual human anatomy, Precise knowledge of human body structure, Even if there is a operational mistake, And practice back and again, No need to worry about wasting resources, For the future growth of medical skills, dying and saving doctors to lay a solid foundation. VR education software makes the original abstract and obscure knowledge become visual and operational, greatly improve the learning effect and interest, the boring knowledge into vivid learning experience, stimulate students eagerly learning enthusiasm, let them take the initiative to roam in the vast ocean of knowledge; vocational training clever use simulation scene, let workers, pilots and other professionals repeatedly practice skills in the virtual environment, greatly shorten the training cycle, significantly improve the practical operation ability, continuously cultivate high-quality professionals for various industries, and meet the urgent needs of social development for professional skills.

Yuan universe has huge commercial potential, virtual land transactions, digital asset creation and sales are extremely hot, triggering an investment boom, attracting a lot of capital, and become one of the most hot and focused topics in the current economic field. However, just as all new things often face many troubles and challenges in the early stage of growth, the metacom is also in deep trouble, and there are many thorny problems to be solved. Some users addicted to it, cannot extricate themselves, gradually ignore the real life, physical and mental health damage, if into a sweet but dangerous and deadly "trap"; privacy and security risks cannot be ignored, personal information in the virtual space as "streaking", vulnerable to lawbreakers; content supervision problems are extremely prominent, the virtual world information is complex and chaotic, bad content is easy to spread, to users, especially young groups. To ensure the healthy and orderly development of the universe, needs the government, enterprises, academia hand in hand, work together to formulate norms, clear entry criteria, content audit criteria, key elements, strengthen the user guidance, make it truly expand human survival dimension, enrich the innovation of the spiritual world carrier, rather than cause a series of social problems "curse", let the universe fully release potential, Truly benefit people.

7.1.6 Value Symbiosis: The Noahs Ark of the Common Future of Mankind

The digital age is like an invisible but tenacious magical bond, like a thin silk thread connecting everything through time and space, tying human destiny together more than ever before. In the depth of globalization interconnected, like a large complex, exquisite of the background, the ecosystem of symbiotic value become the only way to deal with global challenges, towards a better future, just like a human hope Noahs ark, in the era of choppy, dark tide tide for human firm guide direction. In the field of ecological protection, digital technology is like a wise and intelligent "guardian", giving strong energy to environmental monitoring and governance. Satellite remote sensing technology like hanging sky, the "eye", moment overlooking the every inch of the earth, the Internet of things sensor like throughout the earth, sensitive "tentacles", accurately sense the subtle changes of the surrounding environment, their real-time collection of atmosphere, water quality, soil, and other huge amounts of data, build up the earth environment accurate "digital portrait". Through big data analysis, like a detective who has experienced experience and breaks a case to accurately locate the pollution source, the environmental protection department has formulated targeted governance strategies to carry out pollution prevention and control work in a targeted way, so that the earth's environment can be obtained Effective protection; smart grid system uses digital technology to optimize energy distribution, monitor energy consumption in real time, flexibly allocate power resources according to different regions and different periods, reduce waste, improve energy efficiency, help global sustainable energy development, as if carefully installed intelligent "energy steward" for the earth, to make the use of energy more reasonable, efficient and branch.

In the field of public health, the global pandemic, an unprecedented disaster, has impressed people with the urgency and importance of international cooperation. The concept of value symbiosis runs the way through the fight against the epidemic. With the help of digital platforms, national research teams selflessly share key information such as viral gene sequences and clinical treatment experience, Resolutely breaking down the limits of national boundaries, Greatly accelerated the vaccine development process, Many international pharmaceutical companies jointly tackle, Breaking through the technical bottlenecks, Rapid launch of multiple effective vaccines, To inject strong confidence into the global fight against the epidemic, As in the dark boundless night lit a beacon of hope; Telemedicine consultations, across national borders, Experts provide professional and accurate diagnosis and treatment advice for foreign patients online, Effectively relieve the pressure of treatment in areas with tight medical resources, Save countless living lives, Like building a "transnational bridge" of life under the shadow of the epidemic, Let love and hope be transmitted around the world.

At the level of cultural exchange, digital technology is like an invincible sharp sword, breaking the space-time barriers of cultural communication, and promoting the integration and symbiosis of diverse cultures. Social media platforms enable the wide spread of unique folk art and minority cultures from around the world, Africa's passionate inspiration, Japanese elegance and other harmonious songs have swept the world through short videos, To arouse the strong interest and close attention of people with different cultural backgrounds, Enhancing cultural understanding, Make the world culture more colorful and gorgeous; Digital museums and online libraries use advanced technologies to digitize precious cultural relics and massive books, Let people appreciate the splendid treasures of human civilization without leaving their homes, Nourishing the mind, widen the vision, Create a good atmosphere for cultural inheritance and innovation, Like the opening of a never-ending "global cultural feast", Let people enjoy the endless charm of culture.

In the face of global poverty, the digital economy is like a light of hope, bringing new opportunities for developing countries to alleviate poverty and demonstrating the powerful power of value values. E-commerce platform help agricultural products out of the mountains, sold around the world, China's Yunnan fragrant coffee beans, nutritious nuts in Africa become the international market with network marketing, greatly increase farmers income, farmers their hard work has been rich returns; digital finance for small loans for small and micro enterprises, timely help to solve the problem of capital turnover, support the growth of enterprises, create more jobs, drive the vigorous development of local economy, as if for the hope of the "rich beacon", illuminate the road of poverty and prosperity.

However, the road to realize value symbiosis is by no means smooth and smooth, and still faces many obstacles. Trade protectionism has risen from time to time, Some countries have set up digital trade barriers, To ts the free flow of information, and technology, Seriously affecting the coordinated development of the global industrial chain, Like maliciously placing roadblocks on international cooperative "highways", Let the vehicle forward is blocked to slow down; The ethical controversy over digital technology, Issues such as algorithmic discrimination and data abuse, Breaking the bottom line of justice, Caking social unrest, Need to build a sound ethical framework to regulate and restrain, To protect the digital world; The gap between the north and the south remains significant in the digital age, There is a big gap between developing countries and developed countries in digital infrastructure construction and technology application capacity, In urgent need of joint assistance from the international community, Together to close the gap, To make global development more balanced and coordinated. Only countries abandon unilateralism narrow thinking, uphold the open tolerance, win-win cooperation attitude, relying on digital technology to strengthen cooperation, to make value symbiosis concept take root, blossom, lead the human to the bright future, ensure that Noahs ark steady forward, carrying the hope and dream of mankind, ride the waves to the other shore of happiness.

7.2 Digital Technology Revolution: The Force of Negative Entropy Drives Human Progress

The digital technology revolution in this era is like a bright superstar, illuminating the way forward of human society. The negative entropy brought by it is promoting the human society in the sustainable development, value consensus and value innovation with a powerful force.

Sustainable development is a major issue facing the world today, and digital technology plays a vital role in it. Relying on digital technology, the intelligent energy management system realizes the accurate monitoring and efficient control of all links of energy production, transmission and use. Through the in-depth analysis of big data and the accurate algorithm of artificial intelligence, energy distribution can be optimized according to actual demand, greatly improve energy utilization efficiency and effectively reduce energy waste. For example, the smart grid can sense the changes in energy supply and demand in real time and make dynamic adjustments to ensure a stable energy supply while achieving reasonable distribution. Digital technology also contributes to environmental protection. Environmental monitoring sensors and satellite remote sensing can continuously collect massive environmental data, so that we can more accurately grasp the trend of environmental changes, so as to take timely targeted environmental protection measures. The rise of digital education has broken the time and space limitations of traditional education, and enabled knowledge to be widely spread and shared across regional and time barriers. Online education platforms provide a variety of learning resources for people from different regions and backgrounds. No matter you are in a remote mountainous area or a bustling city, you will have the opportunity to use these platforms to improve your knowledge and skills. This has not only effectively narrowed the educational gap, It has promoted social equity and laid a solid foundation for cultivating talents to meet the needs of future social development. Telemedicine and health management have made major breakthroughs, driven by digital technology. Through telemedicine technology, patients can easily get professional medical diagnosis and treatment advice even at home, greatly improving the accessibility of medical services. For those patients living in remote areas, this is undoubtedly a significant medical revolution. At the same time, digital health management devices can monitor people's health status in real time, providing an important basis for disease prevention and personalized treatment.

In terms of value consensus and value innovation, digital technology also shows its unique charm and great potential. The development of quantum technology and block chain technology has brought a new hope for the reconstruction of the trust system. Block chain, with its remarkable features such as decentralization and imtamability, ensures the authenticity and security of data, and provides a strong technical support for the establishment of a more transparent and fair trust mechanism. The rapid progress of quantum technology has built a stronger protective wall for information security, making the trust foundation of the digital world more stable and reliable. With the continuous development of quantum technology and artificial intelligence, people's thinking on values and moral ethics is becoming increasingly deep. The widespread use of AI in various fields prompts us to think seriously about how to ensure that its decisions meet human values and ethical standards and avoid possible adverse consequences. In autonomous driving, for example, we must ensure that AI systems can make the right choices in the face of ethical dilemmas. At the same time, the major breakthroughs in quantum technology have also prompted us to re-examine the existing value system and actively explore the values that adapt to the new era, so as to better cope with various challenges in the future. Digital culture and cultural exchange and mutual learning industry is booming with the powerful boost of digital technology. Digital media, virtual reality, etc. Advanced technology provides a new means and way for cultural dissemination and innovation. The cultures of different countries and regions can have more extensive and in-depth communication and integration through digital platforms, which greatly promotes the diversity and richness of culture. The vigorous development of the digital cultural industry has not only created huge economic value, but also enhanced the mutual understanding and friendship among different ethnic groups, laying a solid cultural foundation for the construction of a community with a shared future for mankind.

In short, the negative entropy brought about by the digital technological revolution is a powerful driving force for the progress of human society. We should fully realize the positive role of digital technology, actively embrace the digital technology revolution with an open mind, constantly explore and innovate, give full play to the advantages of digital technology in sustainable development, value consensus and value innovation, and make concerted efforts to promote the human society toward a better future.

7.2.1 Digital Technology: Leading Human Civilization Towards Sustainable Development

In today's era, digital technology has profoundly changed all aspects of human society with its powerful power. It has not only brought unprecedented convenience to our lives, but also plays a vital role in promoting the journey of human civilization towards sustainable development.

Digital technology has opened up new ways for intelligent energy management and environmental protection. In the face of the global energy crisis and environmental challenges, digital technology realizes the intelligent monitoring, control and optimization of energy systems through the Internet of things, big data and artificial intelligence. Smart meters monitor the electricity consumption of households and enterprises in real time to provide energy saving suggestions for users; smart sensors are installed on energy equipment and facilities to monitor the operation status and energy parameters to improve the reliability and security of the energy system. Big data analysis conducts deep mining of massive energy data to understand energy supply and demand, use efficiency and waste. For example, by analyzing urban traffic flow and energy consumption data, optimizing traffic signal control, reduce vehicle congestion and energy waste, analyzing energy consumption data of industrial enterprises, discover energy saving potential and develop targeted plans. Artificial intelligence realizes the intelligent optimization and control of energy system, and the intelligent energy management system adjusts the production and consumption strategy according to the changes of energy supply and demand and price to realize the optimal distribution and utilization; the smart grid predicts power load through artificial intelligence algorithm to improve stability and reliability and reduce power failure accidents. In terms of environmental protection, sensor technology is installed in environmental media to monitor environmental quality parameters in real time, and transmit data through the Internet of Things, once exceeded Standard automatic early warning, remind to take measures. Satellite remote sensing technology carries out environmental monitoring and evaluation in large areas, and understands land use, forest coverage and water resources distribution information through image and data analysis, so as to provide a basis for environmental protection and ecological restoration. For example, the monitoring of natural disasters will take timely rescue and mitigation measures, and the monitoring of illegal activities will strengthen supervision and enforcement.

The application of digital technology in education cultivates innovative talents for global sustainable development. Online education platform breaks the limitation of time and space and provides rich learning resources. Students can choose courses according to their interest needs, and teachers can conduct teaching management and interactive communication through the platform to improve teaching efficiency and quality. Distance education technology will deliver high-quality education resources to remote and poor areas, narrow the education gap, promote educational equity, and provide convenient learning opportunities for in-service personnel and lifelong learners. Virtual laboratory technology provides a realistic experimental environment to improve the effect and security of experimental teaching. Students can conduct experimental operations through virtual reality and other technologies, which can understand the experimental principle and process more intuitively, and improve their experimental skills and innovation ability. The knowledge-sharing platform promotes knowledge dissemination and innovation, shares experiences, insights and innovations with people through the Internet and social media, conducts exchanges and cooperation, and provides intellectual support for solving global problems.

The development of digital technology in the medical field guarantees the human health and well-being. Telemedicine technology enables patients to enjoy professional medical services in the home or community, reducing the cost and time of medical treatment, and improving the accessibility of medical services. Doctors provide treatment advice and prescriptions to patients through video conferencing and remote diagnosis, serving patients in remote areas, and narrowing the medical gap. Mobile medical technology allows patients to access medical services and health management information anytime and anywhere through mobile devices, including online consultation, appointment registration and inspection reports. The wearable devices monitor health status in real time and provide diagnostic reference for doctors. Electronic medical record technology realizes the digital management and sharing of medical information, and improves the efficiency and quality of medical services. Doctors quickly consult medical record information to understand the medical history, diagnostic results and treatment plan. Doctors in different hospitals can share medical record information for remote consultation and collaborative treatment. The health management platform provides personalized services, and people record health data for evaluation and early warning, obtain health advice and intervention programs, and develop a healthy lifestyle.

Digital technology promotes global cooperation and exchanges, and provides the impetus for the sustainable development of human civilization. It breaks regional and cultural restrictions and makes it easier for people to communicate and cooperate. Countries share experience and technologies for sustainable development through digital platforms to jointly address global challenges. In addressing climate change, we should strengthen cooperation to promote the development and application of clean energy and reduce greenhouse gas emissions. Scientists use digital technologies to cooperate on scientific research projects around the world and explore solutions to global problems. Enterprises cooperate internationally through digital platforms to develop innovative products and services and promote the sustainable development of the global economy. Digital technology also promotes cultural exchanges and integration. The cultures of different countries and regions can exchange and learn from each other through the Internet and social media to enhance mutual understanding and friendship. The digital cultural industry provides new opportunities and platforms for global cultural exchanges and promotes the common progress of human civilization.

In short, digital technology plays an irreplaceable role in promoting the road of human civilization towards sustainable development. We should fully recognize its huge potential, actively promote the deep integration of digital technology and sustainable development, strengthen international cooperation and exchanges, respond to global challenges, and strive to create a better future.

7.2.2 *Digital Technology: A Powerful Engine Driving Value Consensus and Innovation*

In today's era, digital technology has profoundly changed all aspects of human society. From the reshaping of the trust system to the thinking of value concepts, from the extensive application of big data to the vigorous development of the digital culture industry, digital technology is playing a vital role.

Quantum technology and block chain technology in digital technology have brought new hope for the reconstruction of the trust system. As a cutting-edge technology, quantum technology has great potential and influence. Quantum communication uses the characteristics of quantum mechanics, such as quantum entanglement and the non-cloning principle, to realize the safe transmission of information. Compared with traditional communication methods, quantum communication can ensure the absolute security of information in the transmission process, and provide reliable protection for sensitive information and state secrets in financial, military, government and other fields. For example, quantum key distribution technology can generate secure keys for both parties, making it impossible to steal communication content even in the presence of malicious attackers. In addition, the development of quantum computing will also have a profound impact on trust systems. Although quantum computing is still in the early stages of its development, it has attracted much attention. Quantum computing has powerful computing power and can solve complex problems that are difficult to handle by traditional computers in a short time, which may change the pattern of cryptography and make the existing encryption algorithms face challenges. Therefore, we need to study and develop new quantum security technologies in advance to deal with the threats brought by quantum computing, and at the same time to provide new ideas and methods for the reconstruction of trust system.

Block chain technology is a distributed ledger technology, with decentralization, non-tamper, transparency and other characteristics. It plays an important role in the reconstruction of the trust system. First, block chain technology can achieve decentralized trust. In the traditional trust system, it usually needs to rely on a third party to conduct trust endorsement, while block chain technology, through distributed ledger and consensus mechanism, enables each node in the network to jointly maintain the authenticity and integrity of the ledger, without relying on the third party. This decentralized trust mechanism can reduce the cost of trust and improve the efficiency of trust, especially suitable for cross-border payment, supply chain finance and other fields. Secondly, the immutable nature of block chain technology can ensure the authenticity and reliability of data. Once the data is recorded on the block chain, it cannot be tampered with, giving the data on the block chain a high degree of confidence. For example, in the field of food traceability, block chain technology can record the information of the production, processing, transportation and other links of food, and consumers can query the source and quality of food through block chain, so as to improve the trust in food. Finally, the transparency of block chain technology can promote trust building. All transaction records on the block chain are open and transparent, and can be viewed by anyone. This transparency, It can reduce information asymmetry, improve the transparency and fairness of the market, and promote the establishment of trust. For example, in the field of real estate transactions, block chain technology can record the ownership of real estate, transaction history and other information, making the transaction more transparent and secure.

The integration of quantum technology and block chain technology will bring new opportunities for the reconstruction of the trust system. Quantum key distribution technology can provide a more secure encryption method for the block chain, and the transaction information on the block chain needs to be encrypted and protected from tampering and theft. Quantum key distribution technology can generate secure keys that make the encryption on the block chain more reliable. At the same time, the powerful computing power of quantum computing can also provide support for the consensus mechanism of block chain and improve the performance and efficiency of block chain. For example, quantum block chain technology can be applied in the financial sector to achieve more secure and efficient cross-border payments and settlement. Quantum key distribution technology can ensure the safe transmission of payment information, while quantum computing can accelerate the consensus process of block chain and improve the processing speed of payment. In addition, quantum block chain technology can also be applied to the Internet of Things, supply chain and other fields, providing new solutions for the construction of trust systems in these fields.

Quantum technology and artificial intelligence in digital technology have also triggered peoples profound thinking on values and moral ethics. The development of quantum technology has not only brought about technological changes, but also triggered people to think about value ideas. The worldview of quantum mechanics is very different from the traditional classical physics. Quantum mechanics emphasizes microscopic uncertainty and probability in sharp contrast to the traditional deterministic worldview. This change in the world outlook may affect people's values, making people pay more attention to uncertainty and possibilities, and have more respect for individual choice and freedom. In addition, the development of quantum technology may also bring about new values. For example, the powerful computing power of quantum computing may drive the rapid development of science and technology, thus leading to more innovation and progress. Such innovation and progress may encourage people to pay more attention to the value of knowledge and innovation, and to pursue more truth and wisdom.

The development of artificial intelligence has also had a profound impact on value concepts. The emergence of artificial intelligence enables machines to simulate the intelligent behavior of human beings, which causes people to think about the nature and value of human beings. If machines can think and act like humans, what is the unique value of human beings? This is a question worth thinking about deeply. At the same time, the development of artificial intelligence may also bring about new values. For example, artificial intelligence can help humans solve complex problems, improve productivity, and improve the quality of life. The progress of this technology may encourage people to pay more attention to the value of efficiency and efficiency, and to pursue a more convenient and comfortable life. However, the development of AI has also brought some potential risks and challenges. For example, AI may cause unemployment problems and increase social inequality; AI may be used for malicious purposes, such as cyber attacks, wars, etc. These problems require us to seriously think about the development direction and value orientation of artificial intelligence, so that the development of artificial intelligence is in line with the interests and values of human beings.

The combination of quantum technology and artificial intelligence brings us great opportunities, but it also brings a series of moral and ethical problems. First, the development of quantum technology and artificial intelligence may pose a threat to human privacy and security. The powerful computing power of quantum computing may crack the existing encryption algorithms, leading to the disclosure of personal privacy and trade secrets. The widespread use of AI may also collect large amounts of personal data that, if abused, may pose a serious threat to personal privacy and security. Second, the development of quantum technology and artificial intelligence may raise the questions of moral responsibility. If quantum computers and artificial intelligence systems make wrong decisions, who should be responsible? Is it the developer, the user, or the machine itself? This is a question that needs us to think about seriously. Finally, the development of quantum technology and artificial intelligence may have profound effects on the future of mankind. We need to think about how to ensure that the development of quantum technology and artificial intelligence is in the long-term interests of human beings and to avoid uncontrollable risks and consequences.

Big data is also an important part of digital technology, which plays an important role in value consensus and value innovation. The wide application of big data enables us to collect and analyze a large amount of information, so as to better understand the operation law and development trend of human society. In this process, big data also promotes the formation of value consensus. First, big data can help us better understand social needs and public opinion. Through the analysis of social media, Internet search and other data, we can understand the hot issues and needs that people concerned, so as to provide reference for policy formulation and social governance. For example, the government can analyze big data to understand the publics support and opposition to a certain policy, so as to better adjust policies and improve the effectiveness and satisfaction of policies. Secondly, big data can promote exchanges and cooperation between different fields. The cross-field application of big data enables experts and scholars in different fields to analyze and solve problems together, thus promoting the value consensus and cooperation between different fields. For example, in the medical field, big data can help doctors better understand the occurrence and development of diseases, so as to develop more scientific and reasonable treatment plans. At the same time, big data can also promote cooperation in healthcare and other fields, such as information technology. Cooperation in the development of medical big data platform to improve the quality and efficiency of medical services. Finally, big data can promote global cooperation and common development. The global application of big data enables people in different countries and regions to share data and knowledge together, thus promoting global cooperation and common development. For example, in the field of climate change, countries can share climate data and research results and jointly formulate policies and measures to address climate change to achieve sustainable global development.

Big data not only promotes the formation of value consensus, but also provides new opportunities for value innovation. First, big data can drive business model innovation. Through the analysis of user data, enterprises can better understand the user needs and behaviors, so as to develop more personalized and accurate products and services, and improve user satisfaction and loyalty. Second, big data can promote scientific and technological innovation. The analysis and mining of big data can provide new ideas and methods for scientific research and promote the innovative development of science and technology. For example, in biology, big data can help scientists to better understand the function of genes and the mechanism of disease occurrence, and thus develop more effective treatments. Finally, big data can promote innovation in social governance. The application of big data can improve the accuracy and effectiveness of social governance, and realize the modernization of social governance. For example, through the analysis of urban traffic data, the government can optimize the control of traffic lights and alleviate traffic congestion; Through the analysis of social security data, the police can predict the occurrence of criminal events in advance and improve the level of social security.

The development of digital culture and cultural exchange and mutual learning industry is also an important embodiment of digital technology to promote value consensus and value innovation. Digital culture refers to the cultural form with digital technology as the carrier, including digital music, digital film, digital games, digital art and so on. Digital culture has the characteristics and value of innovation and diversity, interactive and participatory, global and sharing. First, digital culture is innovative and diverse. The development of digital technology provides new tools and means for cultural creation, and makes cultural works more colorful. For example, digital music can create various styles of music through electronic synthesizers and music production software; digital movies can use special effects and 3D technology. Secondly, digital culture is interactive and participatory. Digital technology makes the dissemination and consumption of cultural works more convenient, but also provides more opportunities for users to interact and participate. For example, digital games allow players to interact and compete in virtual worlds; digital art allows viewers to participate in the creation and appreciation of art through virtual reality. Finally, digital culture is global and shared. Digital technology breaks down the geographical and language restrictions, making cultural works fast on a global scale Communication and sharing. For example, a popular digital music can spread around the world in an instant, and a wonderful digital movie can attract the attention of a global audience. The global and shared nature of digital culture promotes the communication and integration between different cultures and enriches the diversity of human cultures.

Digital cultural exchange and mutual learning industry refers to the industrial form with digital culture as the core and cultural exchange and mutual learning through digital platforms such as the Internet and mobile Internet. The development of the digital cultural exchange and mutual learning industry is of great significance. First, the digital cultural exchange and mutual learning industry can promote the communication and integration between different cultures. Through digital platforms, cultural works from different countries and regions can be spread and communicated with each other, promoting the understanding and respect between different cultures, and promoting the common development of human culture. For example, Chinese online literature is widely welcomed overseas, promoting the exchange and integration of Chinese culture and the cultures of other countries. Secondly, the digital cultural exchange and mutual learning industry can promote cultural innovation and industrial upgrading. The development of digital technology has brought new opportunities and challenges to cultural creation and dissemination, and promoted cultural innovation and industrial upgrading. For example, the development of the digital cultural industry has promoted the organic integration of the traditional cultural industry and the digital technology, and promoted the digital transformation and upgrading of the traditional cultural industry. Finally, the digital cultural exchange and mutual learning industries can promote economic development and employment growth. Digital culture industry is an emerging industry with huge development potential. Its development can drive the development of related industries and promote economic growth and employment growth for instance. The development of the digital game industry can drive the development of the entire game development, operation, promotion and other related industries, and create a large number of employment machines.

In short, digital technology plays an important role in promoting value consensus and value innovation. From quantum technology and block chain technology for the reconstruction of trust system, to the quantum technology, artificial intelligence value concept and ethical thinking, from the application of big data to the digital culture and cultural exchange of the development of mutual learning industry, digital technology is with its unique charm and powerful force leading the human society towards a new stage of development. We should fully realize the importance of digital technology, actively promote the innovation and application of digital technology, and promote the formation of value consensus and the innovative development of value.

7.3 Entropy Increase and Negative Entropy Balance Under the Digital Technology Revolution

Under the wave of the digital technology revolution, the human society is faced with both the challenges brought by the entropy increase and the opportunities brought by the negative entropy. In order to achieve sustainable development, we need to work hard in value consensus, behavior norms and order reconstruction to achieve the balance of entropy increase and negative entropy.

Value consensus is an important basis for achieving a balance. In terms of international rules, with the globalization of digital technology, it is very important to formulate a unified and effective international digital rules. Different countries and regions differ in the application and development of digital technology, which may lead to trade disputes, data sovereignty disputes and other issues. Therefore, the international community needs to make joint efforts to formulate fair and reasonable international digital trade rules and data protection rules, so as to regulate the development of the global digital economy.

The international social order has also been deeply influenced by the digital technological revolution. The rapid development of digital technology could break the existing international balance of power and trigger new geopolitical competition. At the same time, cyber attacks, false information dissemination and other issues also pose a threat to the stability of the international community. In order to maintain the international social order, countries need to strengthen cooperation to jointly fight cyber crime, regulate behavior in cyberspace, and ensure the peaceful and safe application of digital technologies.

Developing ethical guidelines and norms for digital technology development is the key content of value consensus. The development of digital technology has brought about a series of ethical problems, such as the decision-making fairness of artificial intelligence and the privacy protection of big data. We need to develop clear ethical guidelines that ensure that technology development to human values and ethical standards. At the same time, we should ensure the fairness, transparency and sustainable development of technology, prevent digital technology from exacerbating social inequality, protect personal privacy and data security, and prevent the abuse of personal information.

The concept of behavior is an essential tool in the quest to balance the relentless march of entropy with the infusion of negative entropy. In today's rapidly evolving digital landscape, the importance of strengthening digital literacy education and training cannot be overstated. It is through such educational initiatives that individuals can significantly enhance their cognition and application ability of digital technology.

In the digital age, the necessity for a certain level of digital literacy has become ubiquitous. This encompasses a wide array of skills and awareness, such as the ability to effectively retrieve and evaluate information from the vast sea of data available online. It also includes a robust understanding of digital security, which is crucial in protecting personal information from cyber threats. Furthermore, the ability to navigate network social etiquette is vital for maintaining respectful and productive interactions in digital communities.

Digital literacy is not just about technical skills; it's also about fostering a mindset that appreciates the ethical and social implications of technology use. For instance, understanding the impact of one's digital footprint can lead to more responsible online behavior.

Through comprehensive education and training programs, people can better adapt to the ever-changing demands of the digital society. These programs can range from basic computer skills workshops to advanced cybersecurity courses, ensuring that individuals at all levels of technological proficiency can find the resources they need to thrive.

Moreover, as digital technology becomes increasingly integrated into every aspect of life, from work to personal relationships, the ability to use it effectively and responsibly becomes paramount. Without proper digital literacy, individuals risk falling behind in the professional sphere and may be more susceptible to the negative effects of misinformation, cyberbullying, and privacy breaches.

By investing in digital literacy education and training, society can mitigate the risks associated with the improper use of digital technology. This proactive approach not only empowers individuals to harness the full potential of digital tools but also contributes to a more informed and secure digital society. In essence, digital literacy education is a cornerstone for building a resilient and adaptive community that can navigate the complexities of the digital age with confidence and competence.

Cultivating correct digital values and behavior habits is also crucial. The development of digital technology makes the dissemination of information more rapid and extensive, but it is also easy to lead to the emergence of false information, network violence and other bad phenomena. We need to cultivate peoples correct digital values, advocate civilized and rational network behavior, and jointly create a healthy and harmonious digital environment.

Order reconstruction is an inevitable requirement to realize the entropy increase and negative entropy balance. Promoting international cooperation and governance in digital technology is the key to addressing the global digital challenges. The development of digital technology is a global problem that no country can handle alone. Countries need to strengthen cooperation to jointly formulate strategies and policies for the development of digital technology, share experience and technology, and jointly solve the challenges brought about by digital technology.

To jointly build an open, fair and reasonable digital order is the goal of achieving global digital development. This requires countries to strengthen domestic digital governance on the basis of international rules, establish and improve a sound digital laws and regulations, and protect the rights and interests of consumers. At the same time, innovation should be encouraged to create a favorable policy environment and market environment for the development of digital technology.

In short, the digital technology revolution has brought great challenges and opportunities to the human society. We need to make efforts in value consensus, code of behavior and order reconstruction, to achieve the balance between entropy increase and negative entropy, and to promote the sustainable development of digital technology.

7.4 Value Consensus Building in the Era of Digital Technology

In today's era of rapid development of digital technology, the world is undergoing a profound and extensive change. The wide application of digital technology has greatly changed peoples way of life and social and economic structure, and at the same time has brought new challenges to international rules, international social order, ethics and many other aspects. In order to realize the sustainable development of digital technology, effectively protect personal privacy and data security, it has become a top priority to build a value consensus.

The global development of digital technology makes it difficult for the traditional international rules to adapt to the new situation. In the field of digital trade, its rapid development has made the traditional trade barriers gradually ineffective, but new digital trade barriers are constantly emerging, such as data localization requirements, digital service tax and other issues, which have become the focus of international trade. In order to promote the healthy development of digital trade, countries need to jointly negotiate and formulate fair and reasonable digital trade rules under the framework of the World Trade Organization and other international organizations, so as to eliminate digital trade barriers and ensure the freedom, fairness and openness of digital trade. In terms of cyberspace security, cyber attacks, cyber crimes and other issues have become global threats, and no country can be immune from them. The international community must strengthen cooperation, establish and improve the cyberspace security governance mechanism, strengthen cyber security technology cooperation, jointly fight cyber crimes, and maintain the security and stability in cyberspace. In terms of intellectual property protection, the development of digital technology has also brought new challenges to intellectual property protection. The replication and dissemination of digital content is easier, and the problem of intellectual property infringement has become increasingly serious. The international community needs to strengthen cooperation on intellectual property protection, formulate intellectual property protection rules adapted to the digital age, strengthen copyright protection of digital content, and crack down on intellectual property infringement.

The development of digital technology has also had a profound impact on the international social order. The popularization of digital technology makes the dissemination of information more rapid and extensive, which not only provides convenience for the exchange and cooperation of the international community, but also may cause information confusion and the conflict of public opinion. At the same time, the development of digital technology may aggravate the inequality in the international community, and then affect the stability and harmony of the international community. To maintain the international social order, all countries need to work together to strengthen cooperation in digital governance. On the one hand, all countries should strengthen information management and standardize the information dissemination order. In the digital age, the authenticity and reliability of information are very important. Countries need to strengthen the supervision of Internet information and crack down on the spread of false information and rumors to safeguard the public interests. At the same time, we should also strengthen the management of digital platforms such as social media, regulate user behavior, and prevent cyber violence and malicious attacks. On the other hand, countries need to strengthen the fair distribution of digital technologies and narrow the digital divide. While the development of digital technology brings great development opportunities to some countries and regions, it may also aggravate the inequality in the international community. In order to realize the common development of the international community, countries urgently need to strengthen the cooperation and exchanges in digital technology, promote the popularization and application of digital technology, narrow the digital divide, and ensure that the development results of digital technology can benefit the whole world. In addition, countries need to strengthen international cooperation in digital technologies to jointly address global challenges. The development of digital technology has brought a series of global challenges, such as climate change, environmental pollution and terrorism. Countries need to strengthen the application of digital technologies in these areas, jointly promote global sustainable development, and maintain the stability and harmony of the international community.

The development of digital technology has brought about economic and social changes, but also caused a series of ethical issues. For example, the development of artificial intelligence may lead to unemployment problems, the application of big data may violate personal privacy, and the development of digital technology may also aggravate social inequality. To ensure the healthy development of digital technology, it is essential to develop ethical norms and norms. The formulation of ethical norms and norms should take into account the common interests of human beings and follow the principles of fairness, justice and openness. At the same time, we should also respect human dignity and rights to avoid causing harm to mankind. The sustainable development of technology should also be considered. The development of digital technology should also follow the principle of sustainable development, pay attention to saving resources and protecting the environment, and avoid causing damage to the natural environment. And we should pay attention to the innovation and progress of technology, and promote the continuous development and improvement of digital technology. In addition, social stability and harmony should be considered, and the development of digital technology should follow the principle of social stability and harmony to avoid adverse effects on the society. For example, we should avoid aggravating social inequality, pay attention to social fairness and justice, and promote the harmonious development of society.

In order to achieve the fair, transparent and sustainable development of digital technology, countries need to work together and take a series of measures. First of all, countries should strengthen the research and development and innovation of digital technology, and improve the level and quality of digital technology. The development of digital technology is a process of continuous innovation. Countries need to increase the investment in the research and development of digital technology, cultivate high-quality digital technology personnel, and promote the continuous progress of digital technology. At the same time, countries also need to strengthen the international cooperation in digital technology, jointly overcome the difficulties in the development of digital technology, and realize the common development of digital technology. Second, countries need to strengthen the supervision and governance of digital technology to ensure the safety and reliability of digital technology. The development of digital technology has brought a series of security and risk problems, such as cyber attacks, data leakage, etc. Countries need to establish and improve the supervision mechanism of digital technology, strengthen the security management of digital technology, and ensure the safety and reliability of digital technology. At the same time, countries also need to strengthen the ethical and legal supervision of digital technology, regulate the application and development of digital technology, and avoid the harm caused by digital technology to human beings. Finally, countries need to strengthen the popularization and application of digital technology and promote the fair development of digital technology. The development of digital technology should benefit all mankind, not the privilege of a few. Countries need to increase their investment in the popularization of digital technology Enter, promote the application of digital technology in various fields, improve the penetration rate and application level of digital technology. At the same time, countries also need to strengthen the training and education of digital technology to improve people's digital literacy and technology.

In the digital age, personal privacy and data security have become the focus of public attention. The development of digital technology has made it easier to collect, store and use personal information, which also poses great challenges to personal privacy and data security. To protect personal privacy and data security, countries need to take a series of measures. First, countries need to strengthen the legislative protection of personal privacy and data security. Strict laws and regulations on personal privacy and data security shall be formulated, clarifying rules for the collection, use and protection of personal information, and penalties for violating personal privacy and data security will be increased. Second, countries need to strengthen the technical protection of personal privacy and data security. Advanced encryption technology and access control technology should be adopted to strengthen the protection of personal information and prevent personal information from being stolen and abused. At the same time, countries also need to strengthen the supervision of digital platforms, requiring digital platforms to strengthen the protection of users personal information, to ensure the security of users personal information. Finally, countries need to strengthen the awareness education of personal privacy and data security. Improve people's awareness of personal privacy and data security, let people understand the importance of personal information, master the methods and skills of protecting personal information. At the same time, countries also need to strengthen the education of enterprises and institutions, and require enterprises and institutions to strengthen the protection of users personal information Protect, fulfill the responsibility and righteousness of protecting users personal information.

In short, in the era of digital technology, the construction of value consensus is the key to realize the sustainable development of digital technology. Countries need to work together to reshape international rules, reconstruct the international social order, formulate ethical norms and norms for the development of digital technology, ensure the fair, transparent and sustainable development of technology, and protect personal privacy and data security.

7.5 The Construction of the Code of Conduct in the Digital Age: Moving towards a Harmonious, Civilized and Innovative Global Digital Society

In the era of accelerating globalization and highly developed digital technology, the human society is moving towards a new era of digitalization at an amazing speed. Digital technology, like a powerful torrent, has swept through all corners of the world, profoundly changing people's life style, working mode and the operation mechanism of the society. However, the rapid development of digital technology is like a double-edged sword, bringing us unprecedented convenience and opportunities, but also bringing a series of severe challenges and problems. At this critical historical node, it is particularly important to build the code of conduct in the digital age. It is not only related to the growth and development of individuals, but also related to the harmony, civilization and progress of the whole society, and even the global rules and order.

Digital literacy education and training, as the cornerstone of building the code of conduct in the digital age, is of irreplaceable importance. With the explosive growth of information, how to obtain and process information quickly and accurately has become a major issue faced by people. Through systematic digital literacy education and training, people can learn to use search engines, databases and other tools to efficiently obtain the required information needed. At the same time, it can also master the methods of screening, evaluation and analysis of information, so as to improve the quality and reliability of information. This is of vital importance to the personal study, work and life in the digital age.

In the field of education, digital literacy education and training can help to cultivate students independent learning ability and innovative thinking. With the booming of online education, students can access rich learning resources through the Internet, but they also face the problems of information overload and uneven quality. Students with good digital literacy can effectively screen out valuable learning resources, make personalized learning plans, and use digital tools to display and communicate their learning results. In addition, digital literacy education can also stimulate students innovative thinking and cultivate their ability to use digital technology in innovative practice. For example, students can express their ideas and ideas through programming, multimedia creation and other ways to lay a solid foundation for their future career development. This is also in line with the trend of global education development. All countries can jointly promote the progress of digital literacy education through exchanges and cooperation, and lay the foundation for the cultivation of talents to adapt to the development of the global digital economy. At the same time, international education organizations can develop unified digital literacy education standards, promote the coordinated development of countries in the field of digital education, and ensure that students have a global perspective and cross-cultural communication ability.

In the work field, digital literacy education and training can improve staff work efficiency and competitiveness. In the digital office environment, employees need to master a variety of digital tools and software, such as office software, project management tools, online collaboration platform, etc. Through digital literacy training, more efficient to complete work tasks. At the same time, employees with digital literacy can also better adapt to the digital transformation of the enterprise and contribute more value to the development of the enterprise. For example, in the marketing department of an enterprise, employees can use data analysis tools to understand consumer needs and develop precise marketing strategies, and in the manufacturing department, employees can realize intelligent management of production process through the Internet of Things technology to improve production efficiency and product quality. In the context of global economic integration, the competition of enterprises is no longer limited to the domestic market, but extends to the global scope. Therefore, improving the digital literacy of employees will not only help enterprises to stand out in the domestic market, but also help enterprises to occupy a favorable position in the international market. Organizations such as the International Chamber of Commerce can promote cooperation and exchanges among companies in digital literacy training, share best practices, and improve the digital competitiveness of global companies.

In addition, digital literacy education and training can also help to enhance peoples cyber security awareness and prevention ability. In the digital age, the network security problem is increasingly prominent, network fraud, network attacks, personal information leakage and other incidents are common. Through digital literacy education, people can learn how to protect personal information security, how to prevent network fraud and network attacks and other knowledge, enhance their own network security awareness and prevention ability. For example, people can learn to set strong passwords, such as strong passwords, regularly update software, and not click on unfamiliar links, to avoid becoming victims of cyber crime. At the same time, digital literacy education can also cultivate peoples digital moral concepts, encourage people to abide by laws, regulations and moral norms in cyberspace, and jointly maintain the security and order in cyberspace. On a global scale, countries need to strengthen cooperation to jointly formulate cyber security rules and standards, fight transnational cyber crime and safeguard global cyber security. The International Telecommunication Union and other agencies can play a coordinating role in promoting national cooperation in the field of cyber security and establishing a global cyber security governance system.

Improving people's ability to apply digital technology is the key link to build the code of conduct in the digital age. The rapid development of digital technology has brought great changes to peoples life and work. Only by constantly improving peoples cognition and application ability of digital technology can we better adapt to the development requirements of the digital age.

Improving peoples cognitive ability of digital technology can help people to understand the nature and characteristics of digital technology in more depth. As an emerging technology, digital technology develops at an amazing speed, and its application fields are beyond imagination. By strengthening the study and understanding of digital technology, people can better grasp the development trend and application prospects of digital technology, and lay a solid foundation for the better application of digital technology. For example, people can learn the basic principles and application scenarios of cutting-edge digital technologies such as artificial intelligence, big data and block chain to understand the impact of these technologies on social and economic development, so as to make better use of these technologies in real life and work. In the wave of global scientific and technological innovation, all countries are actively investing in the research and development and application of digital technology, and strengthening international technical exchanges and cooperation, which can promote the common progress of global digital technology. International Science and Technology Cooperation organizations can build a platform to promote cooperation and exchanges between scientists and engineers from various countries in the field of digital technology and jointly overcome technical problems.

Improving people's ability to apply digital technology can enable people to use digital technology more effectively to solve practical problems. Digital technology has been widely used in various fields, such as education, medical care, finance, transportation, etc. By continuously improving the application ability of digital technology, people can make better use of digital technology to improve their work and study efficiency, and create more value for the society. For example, in the medical field, people can use telemedicine technology to realize the sharing of high-quality medical resources and provide more convenient medical services for patients. In the financial field, people can use digital financial tools such as mobile payment and online banking to conduct convenient financial transactions and improve the efficiency and quality of financial services. Globally, the application of digital technology can promote the optimal allocation of resources, narrow the regional development gap, and promote the sustainable development of the global economy. Countries can jointly explore the application of digital technologies in transnational cooperation, such as cross-border e-commerce and digital trade, to inject new impetus into global economic growth. At the same time, the WTO and other institutions can formulate digital trade rules, regulate the order of the global digital economy, and promote the healthy development of digital trade.

Digital technology is an innovation-driven technology, and its development cannot be separated from peoples innovation and creation. By cultivating peoples digital innovation ability and stimulating peoples innovative thinking and creativity, it can inject new vitality into the development of digital technology. For example, people can exercise their digital innovation ability by participating in digital technology innovation competitions, innovation and entrepreneurship activities, and contribute their wisdom and strength to the development of digital technology. In the global innovation ecosystem, countries can strengthen innovation cooperation, jointly overcome difficulties in the development of digital technology, share innovation achievements, and promote the vigorous development of global digital technology innovation. The International innovation cooperation platform can promote national cooperation in the field of digital innovation, encourage transnational innovation cooperation projects, and promote the development of global digital innovation.

Cultivating the correct digital values and behavior habits is the core content of constructing the behavior norms in the digital age. Digital values refer to peoples value judgment and value orientation of digital technology, digital information and digital behavior in the digital environment. Correct digital values play an important guiding role in people's behavior in the digital environment, and can promote people to set up good behavior habits in the digital age, and jointly create a harmonious and civilized digital society.

Cultivating correct digital values helps people to establish correct digital moral concepts. In the digital age, digital moral problems are increasingly prominent, such as network violence, network rumors, network infringement and so on. These problems have not only seriously affected the order and security of cyberspace, but also caused great harm to peoples physical and mental health. By cultivating correct digital values, people can set up correct digital moral concepts, consciously abide by the network moral norms and code of conduct, and jointly create a friendly, harmonious and civilized network environment. For example, people can respect the intellectual property rights of others in cyberspace, do not arbitrarily spread unverified information freely, and do not participate in bad behaviors such as online violence. On a global scale, countries need to work together to strengthen online moral education, formulate unified online moral norms, crack down on online immoral behavior, and maintain good order in global cyberspace. International Internet governance organizations can promote cooperation among countries in building cyber ethics, formulate global cyber ethics norms, and promote civilization and harmony in cyberspace.

Cultivating correct digital values is not only fundamental for establishing a comprehensive understanding of digital security but also crucial for fostering a vibrant culture of digital innovation. In the digital age, where the landscape is fraught with challenges such as personal information security breaches, property theft, and threats to national security, the importance of instilling proper digital values cannot be overstated. These values empower individuals to recognize and mitigate the risks associated with the digital world, such as personal information leaks, network fraud, and cyberattacks, which unfortunately occur with alarming frequency due to the highly developed nature of digital technology.

By internalizing these values, individuals are encouraged to take proactive measures to safeguard their digital presence. This includes strengthening personal information protection practices, being vigilant about the disclosure of sensitive data, and enhancing awareness of potential cyber threats. Moreover, the cultivation of correct digital values extends beyond personal security to encompass a broader responsibility towards the digital community, promoting ethical behavior and responsible use of digital technologies.

In the realm of digital innovation, the role of correct digital values is equally significant. As the global innovation pattern evolves, countries are recognizing the need to encourage digital innovation, create supportive environments for creativity, and strengthen the protection of intellectual property rights. This holistic approach not only fosters the transformation and application of digital innovation achievements but also lays the groundwork for collaborative efforts to advance the global digital economy.

International Innovation Policy Coordination agencies play a pivotal role in this ecosystem by facilitating cooperation and coordination among countries in the development of digital innovation policies. Their efforts are instrumental in promoting the healthy development of the global digital innovation ecosystem, ensuring that innovations are not only technically sound but also socially responsible and sustainable.

Through the cultivation of correct digital values, individuals and nations can work together to secure a safer, more innovative, and prosperous digital future. This collaborative endeavor enhances digital security and innovation, empowers individuals to navigate the digital world with confidence and creativity, and fosters a global environment where digital technology can thrive for the betterment of all. By embracing these values, we can collectively address the challenges of the digital age and harness the full potential of digital technologies for societal progress and economic growth.

In order to build the code of behavior in the digital age, we need to adopt a series of effective ways and methods.

The government should increase the investment and support for digital literacy education and training. Authenticate and evaluate digital literacy education institutions to improve the quality and level of digital literacy education. At the international level, governments can strengthen cooperation to jointly formulate international standards and norms and promote the coordinated development of global digital literacy education. Intergovernmental international organizations can play an important coordinating role in this regard, promoting exchanges and cooperation among countries on digital literacy education policies.

Schools should incorporate digital literacy education into the education system and strengthen the cultivation of students digital literacy. As an important place to train talents, the school has incorporated digital literacy education into the education system. Schools can cultivate students digital literacy by offering digital literacy courses and organizing digital literacy education activities. For example, schools can offer information technology courses, programming courses, multimedia creation courses to improve students digital technology application ability and innovation ability; organize students to participate in digital technology competition and technology innovation activities to stimulate students innovative thinking and creativity. Schools can also strengthen international exchanges and cooperation, carry out digital literacy education programs with foreign schools, and jointly cultivate talents with a global vision and digital literacy. International school cooperation organizations can promote exchanges and cooperation between schools in various countries in digital literacy education, and promote the digital transformation of global education.

Enterprises should strengthen the digital literacy training for employees, and improve their ability of digital technology application and innovation. As the application body of digital technology, enterprises should pay attention to the cultivation of employees digital literacy and strengthen the training of digital literacy for employees. Enterprises can improve the digital literacy of employees through internal training, external training, online learning and other ways. For example, enterprises can invite professional digital technology experts to train employees, introduce the latest digital technology and application cases; organize employees to attend digital technology seminars, industry conferences and other activities to broaden their horizons and ideas. At the same time, enterprises can also improve their digital innovation ability by encouraging employees to put forward innovation suggestions by carrying out digital technology innovation activities, and inject new vitality into the development of enterprises. In the global economic competition, enterprises can strengthen international cooperation, jointly carry out digital technology innovation projects with foreign enterprises, and enhance their international competitiveness. International enterprise cooperation alliances can promote exchanges and cooperation among enterprises in digital literacy training and innovation cooperation, and improve the digital competitiveness of global enterprises.

Social organizations should actively carry out digital literacy education and training activities to improve the digital literacy of the public. Social organizations can improve the digital literacy of the public by holding digital literacy lectures, carrying out digital technology training, and organizing volunteers to participate in the construction of a digital society. For example, social organizations may invite digital technical experts to hold digital literacy lectures for the public to introduce the basic knowledge and application skills of digital technology; conduct digital technology training activities to provide digital technology training services for the public; organize volunteers to participate in the construction of digital society, such as helping the elderly to learn to use digital equipment and participate in the construction of network civilization, so as to improve the publics digital literacy and social responsibility. Social organizations can also strengthen international cooperation and jointly conduct digital literacy education programs with foreign social organizations to promote the promotion of global digital literacy. The international cooperation platform of social organizations can promote the cooperation of social organizations of various countries in digital literacy education, and jointly promote the development of the global digital society.

The media should also play a positive guiding role in strengthening the publicity and promotion of the code of conduct in the digital age. The media can publicize the code of conduct in the digital age through news reports, special programs, public service advertisements and other forms, and guide the public to establish correct digital values and behavior habits. For example, the media can report successful cases of digital technology innovation, stimulate the public enthusiasm for innovation; expose online uncivilized behavior and cyber crimes, and guide the public to comply with online ethics and laws and regulations. On a global scale, the media can strengthen international cooperation to jointly promote the code of conduct in the digital age, and improve the digital literacy and code of conduct awareness of the global public. International media cooperation organizations can promote the cooperation of the media of various countries in promoting the code of conduct in the digital age, and jointly create a good digital social public opinion environment.

In short, building a code of conduct in the digital age is a long-term and arduous task, which requires the joint efforts of the government, schools, enterprises, social organizations and the media. By strengthening digital literacy education and training, improving peoples cognition and application ability of digital technology, and cultivating correct digital values and behavior habits, we can jointly create a harmonious, civilized and safe digital society. In this digital age, each of us should shoulder the responsibility of building the code of conduct in the digital age, actively participate in the construction of the digital society, and contribute our own strength to promote the development of digital technology and the progress of the society. Only in this way can we better enjoy the convenience and opportunities brought by digital technology, realize personal growth and development, promote social harmony, civilization and progress, maintain global digital rules and order, and jointly build a better global digital society.

7.6 Order Reshaping in the Digital Age: A New Journey for Global Cooperation and Equitable Governance

In today's global digital era, digital technology is profoundly reshaping peoples life mode, working mode and the operation mechanism of society. However, with the large-scale application of digital technology, a series of severe global digital challenges have gradually surfaced, making the reconstruction of order become a top priority. Promoting international cooperation and governance in digital technology, jointly addressing global digital challenges, and thus building an open, fair and reasonable international digital order is of self-evident importance to achieving global stability, prosperity and sustainable development.

The development of digital technology has a remarkable global and cross-field characteristics, and in this digital age, no other country can be immune. International cooperation is undoubtedly a key force to promote the development and application of digital technology. At the level of technology research and development, countries can share resources, knowledge and experience through cooperation to accelerate the process of technological innovation. For example, the research and development of cutting-edge digital technologies such as artificial intelligence, big data and block chain require huge capital investment, a large number of high-quality talents and rich data resources. Countries can jointly overcome technical problems and improve their technical level by establishing international cooperative research institutions and carrying out joint projects. At the same time, international cooperation can also promote the standardization and interoperability of technologies, laying a solid foundation for the application of digital technologies on a global scale. In terms of infrastructure construction, digital infrastructure, as the foundation of the development of the digital economy, covers communication networks, data centers, cloud computing platforms, etc. Countries can work together to build digital infrastructure and upgrade global digital connectivity. For example, under the guidance of the Belt and Road initiative, China has carried out extensive and in-depth cooperation with countries along the Belt and Road in the field of digital infrastructure construction, providing strong support for the development of the world economy besides. International cooperation also contributes to the sustainable development of digital infrastructure. In the field of talent training, the rapid development of digital technology requires a large number of high-quality professionals, and countries can jointly cultivate digital technology talents through cooperation. For example, countries can carry out joint educational programs, hold academic exchange activities, and implement internship and training programs to cultivate digital technical talents with an international vision and innovative ability. At the same time, international cooperation can also promote the flow and sharing of talents, and provide a solid talent guarantee for the development of digital technology.

The development of digital technology has also brought a series of new challenges, such as cyber security issues, digital divide phenomenon, and data privacy dilemma, which urgently need the concerted efforts of all countries. In terms of network security, with the popularization of digital technology, network security problems have become increasingly prominent. Network attacks, data leakage, malware and other security incidents have brought incalculable losses to individuals, units and countries. Countries need to strengthen cooperation, jointly formulate cyber security standards and norms, establish a sound cyber security guarantee system, severely crack down on cyber crimes, and make every effort to safeguard global cyber security. For example, countries can strengthen the research and development of cyber security technologies, share cyber security intelligence, and carry out joint cyber security drills through international cooperation, so as to effectively improve their global cyber security protection capabilities. On the issue of the digital divide, the digital divide refers to the gap between different countries, regions and groups in the access, use and benefit of digital technologies. The digital divide has not only affected the balanced development of the global economy, but also further exacerbated social inequality. Countries need to strengthen cooperation to jointly promote the popularization and application of digital technologies and strive to narrow the digital divide. For example, countries could provide digital technical assistance to developing countries through international cooperation, strengthen digital education and training. Improve the digital infrastructure of developing countries and promote the inclusive development of global numbers. In terms of data privacy, with the development of digital technology, data has become an important resource. However, the issue of data privacy has also attracted increasing attention. The leakage and abuse of personal data and other problems bring serious threats to personal rights and interests and security. Countries need to strengthen cooperation, jointly formulate laws and standards for data privacy protection, strengthen data supervision, and effectively protect personal data privacy. For example, countries can establish security mechanisms for cross-border data flow through international cooperation, strengthen the research and development of data security technologies, and improve data privacy protection and water.

To build an open, fair and reasonable international digital order is an ambitious goal of global digital governance and a solid guarantee for promoting the sustainable development of digital technology. In the formulation of international digital rules, international digital rules are the basis of building an international digital order. Countries need to jointly formulate international digital rules through international cooperation and standardize the development and application of digital technology. For example, within the framework of the WTO, the G 20 and other international organizations, countries can conduct in-depth consultations and negotiations on issues such as digital trade, cross-border data flows and intellectual property protection, and formulate fair and reasonable international digital rules. At the same time, the international digital rules also need to closely adapt to the development and change of digital technology, and constantly adjust and improve. In terms of strengthening the construction of international digital governance institutions, international digital governance institutions are the subjects of implementing international digital rules and maintaining the international digital order. Countries need to strengthen the construction of international digital governance institutions and improve the efficiency and impartiality of international digital governance. For example, countries can strengthen the role of the United Nations, the World Intellectual Property Organization and other international organizations in digital governance, establish special digital governance agencies, and strengthen the coordination and cooperation in international digital governance. At the same time, the international digital governance institutions also need to be widely absorbed. The participation of party stakeholders should improve the democracy and transparency of decision-making. Countries need to strengthen cooperation to jointly crack down on digital monopolies and unfair competition and promote fair competition in the digital economy. For example, countries can strengthen digital market supervision through international cooperation, establish digital anti-monopoly mechanisms, regulate the behavior of digital platforms, and protect the rights and interests of consumers. At the same time, countries also need to strengthen the protection of digital intellectual property rights, encourage digital innovation, and create a good market ring for the development of the digital economy.

In order to promote international cooperation and governance in digital technology, jointly respond to the global digital challenges, and build an open, fair and reasonable international digital order, countries need to adopt a series of practical paths and strategies. First, countries need to strengthen policy communication and coordination, and formulate unified digital development strategies and policies. For example, through bilateral and multilateral dialogue mechanisms, countries can conduct in-depth exchanges and consultations on the development of digital technology, digital policies on the digital economy and digital governance, so as to enhance mutual understanding and trust, and coordinate policy positions and actions. At the same time, countries also need to strengthen their cooperation with international organizations, and actively participate in the formulation of international digital rules and the construction of international digital governance institutions. At the same time, countries also need to strengthen the security of digital technology, jointly address challenges such as network security and data privacy, and create a safe and reliable environment for the development of digital technology. Moreover, countries need to promote the flow and sharing of talents, and jointly cultivate and attract digital technology talents. For example, countries can promote the flow and sharing of digital technology talents and improve the overall quality and level of global digital technology talents by establishing international talent exchange mechanisms, conducting joint talent training programs and providing preferential talent policies. At the same time, countries also need to strengthen digital education and training to improve their own countries. The digital literacy and skills of the people provide talent support for the development of the digital economy. Finally, countries need to promote enterprise cooperation and development, and jointly build a global digital economy industrial chain and innovation chain. For example, countries can encourage their own enterprises to cooperate with foreign enterprises to jointly explore the international market and improve their international competition.

To sum up, in the digital era, promoting international cooperation and governance of digital technology, jointly addressing global digital challenges, and building an open, fair and reasonable international digital order is the inevitable choice to achieve global stability, prosperity and sustainable development. Countries need to fully realize the importance and global importance of digital technology, strengthen policy communication and coordination, deepen technological innovation and cooperation, promote the flow and sharing of talents, promote the cooperation and development of enterprises, and jointly make unremitting efforts to build a better digital world. Only through international cooperation and governance can we effectively respond to the challenges brought by digital technology, realize the sustainable development of digital technology, and make greater contribution to the progress and development of social civilization.

Part III
Open a New Era of Human Civilization in the Digital Era

Chapter 8
The Definition of a Value of Civilization

8.1 Digital Technology: A Revolutionary Force Reshaping Our World

In the contemporary landscape, digital technology stands as a beacon of transformation, not merely ushering humanity into a new era but also vividly demonstrating its immense power and far-reaching influence. This realization has been the cornerstone of my extensive research journey, a odyssey that has spanned over a decade, during which I have delved deep into the realms of digital transformation, seeking to understand its multifaceted impacts on society and the world at large.

My journey began with a simple yet profound question: how is digital technology reshaping our world? To answer this, I embarked on a quest that took me beyond the confines of academic literature and into the heart of the digital revolution. I scoured through more than a hundred books, each offering a unique perspective on the digital age. From the early days of computing to the latest advancements in artificial intelligence, I immersed myself in the history, theory, and practice of digital technology.

But reading alone was not enough. I wanted to see the impact of digital technology firsthand, to understand how it was changing lives and societies around the globe. So, I packed my bags and set out on a journey that took me to dozens of countries, spanning continents and cultures. I visited bustling tech hubs in Silicon Valley, where innovation thrives and new technologies are born. I traveled to remote villages in Africa, where mobile phones are bringing education and healthcare to previously isolated communities. I saw the digital divide firsthand, but also the incredible potential of technology to bridge that gap and create a more connected world.

Throughout my travels, I conducted in-depth interviews with a diverse range of people: politicians, technologists, academics, and everyday citizens. I sat down with government officials to discuss digital policy and regulation, seeking to understand the challenges and opportunities faced by policymakers in this rapidly evolving landscape. I spoke with leading technologists, the pioneers and innovators who are

L. Yan, *Value-Based Civilization*, Understanding China,
https://doi.org/10.1007/978-981-96-9490-7_8

driving the digital revolution forward, to gain insights into the latest advancements and future trends. I engaged with academics, scholars who have dedicated their lives to studying the social, economic, and cultural impacts of digital technology, to explore the theoretical underpinnings of this transformation. And perhaps most importantly, I listened to the stories of ordinary people, individuals whose lives have been touched by digital technology in profound and often unexpected ways.

One of the most striking aspects of my research was the sheer diversity of perspectives I encountered. In some places, digital technology is seen as a panacea, a solution to many of the world's most pressing problems. In others, it is viewed with skepticism and even fear, seen as a threat to jobs, privacy, and traditional ways of life. This diversity of opinion highlighted the complex and often contradictory nature of the digital revolution, a revolution that is reshaping our world in ways that are both exciting and challenging.

As I delved deeper into my research, I began to see patterns and themes emerge. One of the most significant is the idea of a "Human Value-Based Civilization," a concept that encapsulates the potential of digital technology to create a more just, transparent, and sustainable world. This vision is not utopian; it is grounded in the real-world experiences and insights I gathered during my research. It is a vision that recognizes the power of digital technology to empower individuals, connect communities, and drive economic growth, but also acknowledges the need for responsible and ethical use of this powerful tool.

The journey was not without its challenges. Navigating the vast and often complex landscape of digital technology required a deep understanding of both the technical and social aspects of this revolution. It required me to constantly learn and adapt, to be open to new ideas and perspectives. But the rewards were immense. I gained a profound appreciation for the transformative power of digital technology, and a deep understanding of its potential to shape our future.

One of the most powerful moments of my research came during a visit to a rural village in India. I met a young girl who, thanks to a mobile education app, was able to access quality education for the first time. Her eyes lit up as she showed me her progress, proud of the new skills she had learned. This moment encapsulated the true potential of digital technology: the ability to empower individuals and communities, to provide opportunities that were previously out of reach.

But my research also highlighted the challenges and risks associated with the digital revolution. Issues such as data privacy, cybersecurity, and the digital divide are pressing concerns that need to be addressed if we are to fully realize the potential of digital technology. These challenges require a collaborative approach, involving governments, businesses, and civil society working together to create a digital future that is inclusive and sustainable.

As I reflect on my research journey, I am struck by the incredible pace of change. The digital revolution is unfolding at a speed that is unprecedented in human history. This rapid pace of change presents both opportunities and challenges. It offers the potential for unprecedented progress and innovation, but also requires us to be vigilant and proactive in addressing the risks and ethical dilemmas that arise.

Looking ahead, the future of digital technology is filled with possibilities. Artificial intelligence, blockchain, and the Internet of Things are just a few of the technologies that are set to transform our world in ways we are only beginning to imagine. But as we move forward, it is crucial that we do so with a clear vision and a strong set of values. The "Human Value-Based Civilization" is not just a goal to be achieved; it is a guiding principle that should inform our actions and decisions as we navigate the challenges and opportunities of the digital age.

In conclusion, my research journey has been a profound and transformative experience. It has taken me around the world, introduced me to incredible people, and provided me with a deep and nuanced understanding of the digital revolution. I am convinced that digital technology has the potential to create a better, more equitable world, but only if we are willing to harness its power responsibly and ethically. As we stand on the precipice of this new era, it is up to all of us—policymakers, technologists, academics, and citizens alike—to navigate the challenges and seize the opportunities presented by this new era. By doing so, we can ensure that the "Human Value-Based Civilization" is one that benefits all of humanity, fostering a future of shared prosperity and sustainable development. Together, we can shape a digital future that is not only technologically advanced but also socially just and environmentally sustainable. The journey is far from over, but I am optimistic about the path ahead. The digital revolution is here, and it is up to us to make the most of it.

8.2 Value-Based Civilization: The Civilizational Paradigm Under Digital Technological Revolution

What is a Value-Based Civilization? Human Value-Based Civilization is the process of forming a credible consensus on value and reducing entropy increase, driven by digital technology. Through the establishment of value consensus mechanism and the establishment of value order and the construction of governance order, human beings can eliminate the influence of huge dissipative system and form a new Value-Based Civilization paradigm.

"Human Value-Based Civilization" is a new concept and innovation initiated by me and proposed for the first time, which is a new and higher stage of human civilization.

Value-Based Civilization (Human Value-Based Civilization), an innovative theoretical system first proposed by me in 2022, centers on leveraging digital technology clusters such as artificial intelligence, big data, blockchain, and digital twins as driving forces. Through the establishment of value consensus mechanisms, value orders, and governance systems, it aims to address the entropy crisis in human society and construct a new sustainable civilizational paradigm. This theory breaks through the linear thinking of traditional civilizational evolution, integrating technological revolution with value reconstruction, and marks humanity's entry into a new era dominated by digital technologies.

This new paradigm represents humanity's first proactive design of civilizational evolution, offering a path to address entropy crises and achieve long-term sustainability through digital innovation.

"Value-Based Civilization" is a Comprehensive Framework for Human Progress. Value-Based Civilization is a profound concept rooted in the function model of human civilization, where technology and value cognition emerge as pivotal variables driving the evolution of human societies. Within this model, technology acts as the independent variable, constantly evolving and shaping the landscape of human existence. Each stage of technological advancement corresponds to a specific value law, which serves as the dependent variable, adapting and transforming in response to technological changes. However, amidst this dynamic interplay, value itself remains a constant, an enduring force that has woven its way through the tapestry of human history, serving as the core power source that generates the civilization function value of each era.

Human civilization, a testament to human progress, is characterized by continuous revolution and advancement. The pursuit of Value-Based Civilization represents the pinnacle of this progress, encompassing not only technological advancements but also moral, cultural, intellectual, and artistic achievements that collectively constitute the essence of the human soul. This pursuit has always been a fundamental driving force propelling the advancement of human civilization.

As we aspire towards the realization of human Value-Based Civilization, we anticipate the emergence of a new world order characterized by cooperation mechanisms, value-based consensus, and innovative systems. This transformation requires the collective efforts of countless individuals, united by the vision of a global community with a shared future for mankind. Such a vision necessitates collaborative efforts to forge consensus, emphasizing the importance of mutual respect, inclusivity, cooperation, and learning among different groups and nations. These new changes must be grounded in the concept of a global community, ensuring that human civilization and progress are not only realized but also sustained, ushering in a new era of shared destiny.

Value-based civilization also offers a compelling answer to the question of humanity's direction in the midst of the digital technological revolution. It emphasizes the establishment of both value order and governance order. The value order delineates the code of conduct for various scenarios, such as the insistence on real data and rigorous methods in the scientific research field. Meanwhile, the governance order ensures the implementation of this code through regulatory constraints and industry standard supervision. The synergy between these two orders fosters an orderly environment for development, propelling mankind towards new heights of progress.

In essence, "human value-based civilization" is a holistic integration of digital technology innovation, value consensus cohesion, and the dynamics of entropy increase and confrontation. It represents a new paradigm of human civilization, where technological advancements are harmoniously aligned with ethical, cultural, and intellectual values, creating a vibrant and sustainable future for all of humanity. This vision transcends mere technological progress, embracing a broader understanding of human flourishing that encompasses the full spectrum of human experience and potential.

I firmly believe in the transformative power of the "value-based civilization" concept, an innovative idea that I anticipate will make profound contributions to our understanding of human civilization and drive significant social progress. This concept, deeply rooted in the function model of human civilization, emphasizes the critical interplay between technology and value cognition as the driving forces behind our societal evolution.

In my perspective, technology, as the ever-progressing independent variable, is continuously reshaping the landscape of our lives. Each technological breakthrough brings new possibilities and challenges, demanding a reevaluation of our values and societal structures. Meanwhile, value cognition, the dependent variable, evolves in response to these technological advancements, guiding our ethical considerations and societal norms.

The "value-based civilization" concept transcends traditional models by integrating digital technology innovation, value consensus cohesion, and the intricate dynamics of entropy increase and confrontation. This holistic approach creates a synergy that addresses the multifaceted challenges of our time, from environmental sustainability to social equity.

I am convinced that this concept will play a historic role in elevating humanity to unprecedented levels of achievement and fulfillment. It offers a roadmap for navigating the complexities of the modern world, ensuring that our technological advancements are aligned with our deepest values and aspirations. My sincere hope is that this vision will be widely embraced and actively pursued, leading us toward a future that is not only technologically advanced but also morally and socially enriched.

8.3 The Paradigm Innovation of Civilization Spawned by Technological Revolution

The explosive and relentless evolution of digital technology is not merely a fleeting trend but a profound force that is fundamentally reshaping the underlying logic of human civilization. This transformation is a radical departure from the historical paradigms that have governed human progress. Unlike the "survival logic" that characterized agricultural civilization, where the primary focus was on subsistence and basic needs, or the "growth logic" of industrial civilization, which emphasized production and economic expansion, or even the "efficiency logic" of the informational civilization, which prioritized speed and optimization, the emerging Value-Based Civilization introduces a revolutionary "symbiotic logic" paradigm.

This new paradigm is driven by a cluster of advanced digital technologies, including artificial intelligence (AI), big data, blockchain, and more. These technologies do not operate in isolation but converge to create a synergistic effect, breaking through the linear and often incremental evolutionary path of traditional civilizations. The result is a leap in civilizational form, facilitated by a "technology-value double helix" mechanism that intertwines technological innovation with value creation.

8.3.1 Technical Layer: Transcending Boundaries

At the technical layer, innovations such as digital twins and quantum computing are pushing the boundaries of what is possible. Digital twins, for instance, create virtual replicas of physical entities, allowing for real-time monitoring, analysis, and optimization. Quantum computing, with its immense processing power, transcends the physical limitations of traditional computing, enabling humanity to tackle complex problems that were once insurmountable. For the first time in history, humanity has the tools to proactively shape the direction of civilizational evolution rather than merely reacting to it.

8.3.2 Value Layer: Embedding Ethical Principles

The value layer of this new paradigm is equally transformative. Consensus mechanisms, such as those enabled by blockchain technology, embed ethical principles into the very fabric of digital interactions. Concepts like "entropy reduction consensus" and "technology for good" become integral to decision-making processes, ensuring that technological advancements are aligned with societal well-being. This integration of values into technology creates a foundation for a more equitable and sustainable future.

8.3.3 Institutional Layer: Enforcing Value Consensus

At the institutional layer, smart contracts and decentralized autonomous organizations (DAOs) play a pivotal role. These technologies convert value consensus into enforceable rules, creating a closed-loop governance system characterized by "consensus co-construction—automatic rule execution—real-time effect feedback." This system ensures that the values embedded in the technical and value layers are not merely aspirational but are translated into tangible actions and outcomes.

8.3.4 Real-World Applications: A Glimpse of the Future

The paradigmatic innovation of Value-Based Civilization is already demonstrating its unique advantages in addressing some of the most pressing global challenges. For example, the European Union's "Climate Smart Contract" is a groundbreaking initiative that automatically executes carbon quota trading. This system, which covers 23,000 enterprises, is projected to reduce emissions by a staggering 120 million tons in 2024, showcasing the power of digital technology in addressing climate change.

In China, the "East-to-West Data Transfer" project is another testament to the potential of this new paradigm. By using AI scheduling to match computing demands in the eastern part of the country with clean energy sources in the west, the project has successfully lowered the Power Usage Effectiveness (PUE) of western data centers to below 1.25. This not only enhances efficiency but also contributes to sustainable development by reducing the carbon footprint of data centers.

8.3.5 A New Era of Proactive Design

Value-Based Civilization marks a significant shift in humanity's approach to technological change. No longer are we merely passive adaptors to the forces of technology; instead, we are becoming proactive designers of our own civilizational evolution. This transition offers a new path to address the entropy crises that threaten our planet and to achieve sustainable development.

The "technology-value double helix" mechanism is at the heart of this transformation. It ensures that technological advancements are not pursued in a vacuum but are closely linked to value creation, ethical considerations, and societal well-being. This holistic approach is essential for navigating the complex challenges of the twenty-first century.

8.3.6 Conclusion: A Brighter, More Sustainable Future

In conclusion, the emergence of value-based civilization, driven by the revolutionary force of digital technology, represents a pivotal moment in human history. It is a call to action for individuals, communities, and nations to embrace this new paradigm and to work collaboratively towards a future that is not only technologically advanced but also socially just and environmentally sustainable.

The journey ahead is undoubtedly challenging, but the potential rewards are immense. By harnessing the power of digital technology and aligning it with our deepest values, we can forge a path towards a brighter, more sustainable future for all of humanity. The dawn of Value-Based Civilization is upon us, and it is up to us to seize this moment and shape the world we want to live in.

8.4 The Evolutionary Logic of Civilization Under Technological Revolution: A Symphony of Technological Revolutions and Value Transformations

The grand tapestry of human civilization, woven over millennia, has been fundamentally driven by successive technological revolutions, each leading to a paradigmatic leap in our societal structures and value systems. From the dawn of Agricultural Civilization to the emerging era of Value-Based Civilization, we witness a remarkable journey of transformation, where technology and values intertwine to shape the course of human history.

8.4.1 Agricultural Civilization: The Foundations of Collectivism

In the cradle of Agricultural Civilization, bronze and iron technologies emerged as the catalysts for a new way of life. This era was characterized by a collectivist value system, deeply rooted in land worship and patriarchal systems. The land, a symbol of sustenance and survival, became the central axis around which societal values revolved. Centralized governance, with its emphasis on hierarchical structures, became the primary mode of administration, adhering to a "survival logic" that was imperative in addressing the resource constraints of the time. The collective effort to till the land, harvest crops, and sustain communities fostered a sense of unity and interdependence, laying the groundwork for the societal norms and values that would define this era.

8.4.2 Industrial Civilization: The Rise of Individualism and Nation-States

The rumble of steam engines and the glow of electrical lights marked the dawn of Industrial Civilization, a period propelled by revolutionary advancements in steam and electrical technologies. This era witnessed a significant shift in value systems, with capital accumulation and individualistic values taking center stage. The rise of the nation-state as a governance model reflected the growing importance of national boundaries and economic interests. Industrial Civilization adopted a "growth logic," driven by the relentless pursuit of productivity and economic expansion. The factory system, urbanization, and the proliferation of consumer goods transformed societal structures, fostering a culture of individualism and competition. The industrial revolution not only reshaped the economic landscape but also redefined the social fabric, setting the stage for the next phase of human evolution.

8.4.3 Informational Civilization: The Age of Information Sovereignty and Pluralism

As we transitioned into the Informational Civilization, internet and biotechnologies became the new frontier of innovation. This era reshaped value systems through the concepts of information sovereignty and pluralistic inclusivity. The platform economy emerged, characterized by digital platforms that facilitated unprecedented levels of connectivity and interaction. Global governance became more fragmented, reflecting the decentralized nature of information flow and the rise of non-state actors. Informational Civilization was dominated by an "efficiency logic," where the rapid exchange of information and the optimization of processes became paramount. The democratization of information and the proliferation of digital platforms empowered individuals and communities, fostering a more inclusive and diverse societal landscape. However, this era also brought forth new challenges, such as data privacy and digital inequality, necessitating a reevaluation of our value systems and governance models.

8.4.4 Value-Based Civilization: The Emergence of a Symbiotic Paradigm

In the nascent era of Value-Based Civilization, digital technology clusters such as Artificial Intelligence (AI), big data, and blockchain are revolutionizing our understanding of value and governance. This era establishes an entropy reduction consensus, aiming to create order and coherence in an increasingly complex world. The symbiotic value system, characterized by mutualism and interdependence, is at the heart of this new civilization. Smart contracts and Decentralized Autonomous Organizations (DAOs) are the cornerstones of this era, enabling a new "symbiotic logic" paradigm where trust, transparency, and collaboration are paramount. Value-Based Civilization represents a shift from mere efficiency to meaningful collaboration, from isolated growth to sustainable development. It is an era where the proactive design of evolutionary directions becomes possible, thanks to the transformative power of digital technologies.

8.4.5 The Synergistic Mechanism of Technology and Value Reconstruction

This evolutionary process underscores a profound synergistic mechanism between technology-driven value reconstruction and institutional innovation. Agricultural technologies shaped the value of land worship, embedding the importance of community and survival in our collective consciousness. Industrial revolutions spawned capital ethics, emphasizing individualism, competition, and economic growth. Informational revolutions triggered efficiency revolutions, prioritizing speed, connectivity, and inclusivity. Now, digital revolutions are propelling human civilization toward a new era, where the proactive design of evolutionary directions is within our reach.

8.4.6 The Road Ahead: Embracing the Value-Based Civilization

As we stand on the precipice of value-based civilization, the challenges and opportunities are immense. The digital technologies that define this era hold the potential to address some of the most pressing issues of our time, from climate change to social inequality. However, the path forward requires a thoughtful and inclusive approach, one that balances technological innovation with ethical considerations and societal well-being.

8.4.7 Conclusion: A New Era of Human Potential

The evolution of human civilization, from agricultural to value-based, is a testament to the power of technological revolutions and value transformations. Each era has left an indelible mark on our societal structures, shaping our understanding of the world and our place in it. As we embrace the era of value-based civilization, we have the opportunity to harness the potential of digital technologies to create a more equitable, sustainable, and harmonious world. The journey is ongoing, and the future is bright with the promise of human ingenuity and collaboration. Together, we can navigate the complexities of this new era, ensuring that the advancements of today lay the foundation for a better tomorrow.

Chapter 9
Ecological System of Value and Civilization

Amid the tumultuous and ever-accelerating currents of the digital revolution, a paradigm shift has been meticulously orchestrated by the principles of value-based human civilization, giving rise to a highly sophisticated and intricate civilizational ecosystem. This advanced framework is not merely a collection of disparate elements but a harmonious symphony of socioeconomic structures, deeply ingrained humanistic ethos, and delicate ecological matrices, all interwoven through a tapestry of dynamic interdependencies.

These interdependencies are the lifeblood of this evolving ecosystem, ensuring that each component functions not in isolation but in concert with others, creating a resilient and adaptive system capable of navigating the complexities of the modern world. The result is a collective propulsion of humanity toward a horizon of unprecedented civilizational evolution, where the boundaries of what is possible are continually being redefined.

The following analysis delves into the heart of this transformative ecosystem, systematically examining the core mechanisms that act as the engines of change. It explores how technological advancements are seamlessly integrated with ethical considerations, how economic growth is balanced with social equity, and how ecological sustainability is prioritized in the face of industrial progress.

This exploration reveals a vision of the future where the digital revolution is not just a technological phenomenon but a catalyst for holistic human development. It underscores the importance of values in shaping the trajectory of our civilization, ensuring that as we advance, we do so with a sense of purpose, responsibility, and a commitment to the well-being of all humanity and the planet we call home.

In essence, the analysis presents a roadmap for understanding and navigating the complexities of this new era, highlighting the pivotal role of value-based human civilization in fostering a sustainable, inclusive, and prosperous future for all. It is a call to action, inviting us to embrace the challenges and opportunities of the digital age with wisdom, innovation, and a deep-seated respect for the interconnectedness of our world.

L. Yan, *Value-Based Civilization*, Understanding China,
https://doi.org/10.1007/978-981-96-9490-7_9

9.1 New Productivity and Value Factors of Production

In today's changing digital age, a profound change like a surging tide is reshaping every inch of human society with an unprecedented majestic momentum. Among them, the most important and important key elements are the strong rise of the new productive forces and the new transformation of the closely accompanying value production factors. This transformative force, just like a huge stone put into the calm lake, has not only completely overturned the traditional economic development model of inheritance for thousands of years, but also triggered a series of interlocking and far-reaching chain reactions in every corner of the world like ripples.

9.1.1 New Productivity Characteristics Driven by Digital Technology

Digital technology, the quintessential cornerstone of the new era, wields its magic like a sorcerer, infusing the productivity of yesteryears with unimaginable, extraordinary attributes. It breaks the chains of traditional models, much like a beast unshackled, and reveals a vibrant vitality and infinite potential that is both astonishing and profound.

9.1.1.1 A Quantum Leap in Production Efficiency

At the forefront of intelligent manufacturing, the convergence of industrial Internet, artificial intelligence, and big data analysis technology is a match made in heaven. This deep integration is revolutionizing traditional factories, transforming them into highly efficient production centers at an unprecedented speed. The robots on the production line have evolved from mere mechanical executors of simple, repetitive tasks to precision operators with "intelligent brains." They act as keen observers, collecting real-time data on various complex aspects of the production process—such as temperature, pressure, and the accuracy of part dimensions—and then using built-in electronic algorithms to conduct fast and accurate analyses. Like experienced craftsmen, they precisely adjust production parameters, ensuring that every minute link of the production line is dynamically optimized according to actual working conditions. This results in a significant reduction in the product manufacturing cycle, achieving efficiency gains that are nothing short of astonishing.

9.1.1.2 Automobile Manufacturers: Champions of the Revolution

Automobile manufacturers are undoubtedly the poster children for this revolutionary transformation. In the past, switching production lines to accommodate different models was a time-consuming and labor-intensive endeavor, often requiring extensive manpower, resources, and time for complicated equipment debugging and mold replacement. The costs were high, and the efficiency was abysmally low. However, times have changed. Thanks to digital technology, enterprises now possess the magical "Transformers" power of a flexible production system. They can adjust production tasks in the blink of an eye to meet changing market demands and seamlessly connect the assembly work of different models. The benefits of this transformation are clear: production costs have plummeted, and enterprises can now respond to the increasingly diverse and personalized needs of consumers with lightning speed, significantly enhancing their market competitiveness.

9.1.1.3 The Electronics Manufacturing Sector: A Legend of Efficiency

The field of electronic products manufacturing also tells a remarkable tale of efficiency improvement. Take mobile phone production as an example. Guided by the "wisdom lighthouse" of big data, automation equipment operates with the precision of an archer, controlling the temperature and time of each solder joint to ensure welding quality that is as stable as Mount Tai. This naturally leads to a significant reduction in the defective rate. Meanwhile, the intelligent production scheduling system orchestrates the entire process—from raw material procurement to parts processing and complete machine assembly—with exquisite precision. The whole production process runs like a set of high-speed, precisely meshed gears, achieving remarkable efficiency and smoothness. Such synergies allow new phones to be brought to market at a rapid pace, accurately satisfying consumers' eagerness for the latest technology products.**

9.1.1.4 Revolutionary Expansion of the Role Range of Productive Forces

The widespread adoption and application of virtual offices and remote collaboration software have sparked a quiet yet world-shaking revolution. This revolution has decisively broken the long-standing shackles of region and time, allowing global talent resources to converge like rivers flowing into the sea, boldly ushering in a new era of cross-regional collaborative innovation.

In the dynamic field of software development, the landscape has undergone a dramatic transformation. Previously, team members were confined to physical spaces, often crammed into the same office like sardines, relying on face-to-face communication to ensure the smooth progress of projects. However, with the maturation and enhancement of online collaboration platforms, this paradigm has shifted. Now, programmers scattered across the globe—from bustling international metropolises to remote, tranquil border towns—can write, test, and optimize code simultaneously, supported by stable and reliable network connections. The real-time sharing of code repositories, akin to a "treasure house of wisdom," enables them to exchange insights and experiences in solving technical problems at any moment. They skillfully navigate the challenge of time differences, achieving efficient development work around the clock. This new collaboration mode not only accelerates project progress like rocket fuel but also endows enterprises with a sort of "clairvoyance" and "clairaudience," allowing them to tap into the wisdom of the world's best technical personnel, regardless of their geographical location, and continuously inject fresh vitality into product innovation.

Similarly, in the creative and inspiring realm of design, remote collaboration shines brightly, demonstrating significant advantages. Designers now possess a magical pocket, enabling them to freely share design materials through the "infinite space" of the cloud. High-definition video conference software allows them to present design schemes in real time, facilitating face-to-face communication and feedback with customers and team members. For instance, a renowned brand based in Europe's fashion capital can leverage digital technology to collaborate closely with fabric suppliers and talented designers in Asia, jointly creating the latest trends of the season. During this process, design inspiration behaves like an unbound spirit, dancing freely across the globe, while the production process is coordinated with precision, ensuring that new clothing can be launched to the market promptly and with high quality, accurately meeting the stringent demands of the rapidly evolving fashion landscape.

Furthermore, in the challenging and unknown territory of scientific research, international cooperation projects sail forward like ships powered by the strong winds of remote collaboration platforms. Research teams from different countries, like-minded explorers, converge around major issues such as gene editing and quantum computing. By sharing experimental data as "puzzle pieces" and remotely operating experimental equipment with "wisdom," they achieve seamless docking of knowledge and technology, effectively building a "wisdom bridge" that spans borders. This collaborative innovation model, transcending the boundaries of time and space, has significantly accelerated the progress of scientific research and is poised to drive breakthroughs in numerous key fields, heralding a future rich with possibilities and advancements.

9.1.2 The Profound Change of Value Production Factors in the Context of New Productive Force

With the vigorous ascent of new productive forces, akin to an unstoppable surging tide, the very factors of value production have undergone a profound and fundamental metamorphosis. This epochal shift is characterized by the intricate interweaving, collisions, and mergers of traditional factors with emerging elements, collectively crafting a new paradigm brimming with infinite possibilities.

9.1.2.1 Data: The Ascendant Key and Core Element

In this era marked by the tumultuous digital tide, data has ascended to the status of the most dazzling North Star in the night sky of value creation. It has become a pivotal and indispensable core element, its immense value akin to treasure buried deep underground, now gradually unearthed and unleashed through the confluence of human wisdom and technological prowess.

9.1.2.2 Internet Enterprises: Vanguard Warriors in the Data Mining Frontier

Internet enterprises stand at the forefront of this data value mining revolution, akin to vanguard warriors equipped with unparalleled vision. They silently and meticulously collect vast troves of data through every gentle click, casual browse, and curious search conducted on their platforms. These seemingly chaotic data streams, when subjected to sophisticated and magical arrays of algorithmic analysis, yield profound insights into consumer preferences, demand trends, and hidden pain points.

E-commerce giants, for instance, meticulously construct finely detailed and life-like user portraits based on multi-dimensional data such as users' historical "footprints," browsing durations on product details pages, and "wish lists" of favorite content. This process is akin to tailoring a "mirror" of the mind for each user, enabling e-commerce platforms to deliver highly accurate and personalized product recommendations and promotional activities. This strategic move not only acts like a magic wand, significantly boosting user purchase conversion rates, but also uncovers new gold mines, opening up fresh avenues for profit growth. Advertisers, ever keen to seize business opportunities, invest heavily in these precise traffic entry points, eagerly anticipating accurate access to their target customer base.

9.1.2.3 Financial Institutions: Leveraging Data for Risk Assessment and Loan Approval

Financial institutions, too, are increasingly reliant on data, particularly in critical areas such as risk assessment and loan approval. Big data analysis serves as a magical anchor in these life-and-death processes, integrating multi-dimensional data such as users' bank flow "capital context," credit record "credit labels," consumer behavior data, and social media information. This comprehensive approach empowers financial institutions with a seemingly magical "mind-reading" ability, enabling them to use machine learning models to accurately assess credit risks. This transformation not only accelerates and enhances the efficiency and accuracy of the loan approval process but also significantly reduces bad debt rates. Moreover, it opens a door of hope for small and micro businesses and self-employed individuals who struggle under traditional credit evaluation systems, providing them with valuable financing opportunities.

9.1.2.4 Healthcare: Data as a Lifesaver

The value of data in healthcare, a realm where life and death hang in the balance, is growing exponentially. With the widespread adoption of electronic medical record systems and the extensive use of wearable medical devices, massive volumes of patient health data are converging into a vast ocean of information. When analyzed through the "intelligence filter" of artificial intelligence algorithms, these data become invaluable assets, acting as a doctor's right-hand man in assisting disease diagnosis, predicting disease risks, and optimizing treatment plans.

For example, by analyzing extensive blood glucose fluctuation data, diet and exercise records, and medication feedback, medical teams can develop more personalized and effective treatment plans with a "foresight eye" that transcends time and space. This approach not only significantly improves patient rehabilitation effects but also provides fertile soil for medical research and development, fueling advancements that save lives and enhance the quality of healthcare worldwide.

9.1.2.5 Knowledge and Technology Innovation: The Bedrock of Core Competitiveness

In the high-tech industry, a realm brimming with both opportunities and challenges, knowledge and technology innovation stand undeniably as the cornerstone of enterprise competitiveness. These elements enable companies to not only establish a firm foothold in the market but also to navigate the winds of change and generate significant value.

The semiconductor industry, a luminous constellation in this vast star sea of technology, epitomizes this reality. It resembles an unending technological marathon, where the pace of innovation is dizzyingly rapid and the investment in research and development is astronomical. Each minor breakthrough in this field has the potential to trigger a monumental leap in the performance of electronic products, much like the butterfly effect. The world's top semiconductor companies, akin to a cadre of relentless explorers, invest billions annually into chip technology research and development. Their journey spans from the foundational research of basic materials to the optimization of chip design architectures, and the conquest of advanced manufacturing technologies. Each link in this chain is a testament to the collective wisdom and sweat of scientific talent. The breakthroughs in 7 nm and 5 nm processes, for instance, have injected superpower into electronic products, enhancing chip performance and significantly reducing power consumption. This has directly propelled the advancement of smartphones, high-performance computers, and artificial intelligence servers, thereby injecting a strong impetus into the global electronic information industry and creating immense economic value.

Similarly, biotechnology emerges as a new continent, rich in mystery and hope. The development of gene editing technology, like the dawn breaking over a night sky, brings the promise of overcoming many challenging diseases. Research teams, acting as fearless warriors, pour their energies into studying the relationship between gene sequences and diseases, deciphering the code of life, and developing precise gene editing tools aimed at repairing disease-causing genes and preventing genetic diseases. For instance, in the case of hereditary cancers, genetic testing and gene editing technology can intervene precisely, dismantling the "time bombs" of life and potentially stopping diseases at their root. This transformation not only has the potential to alter the medical model but also to give rise to a massive gene therapy industry, creating value beyond imagination for society.

Moreover, the innovation of new energy technologies commands global attention. In the face of escalating environmental problems, researchers, like heroes on a mission, are dedicated to developing efficient solar cells, large-capacity wind power generation equipment, and advanced energy storage technologies. As these technologies continue to break new ground, the proportion of clean energy in the energy structure is steadily increasing, much like bamboo shoots after a rain, laying a solid foundation for achieving global carbon emission reduction targets and promoting sustainable development.

9.1.2.6 New Talent: The linchpin of Value Creation

In the era of new productive forces and emerging value production factors, the demand for compound talents—individuals who possess both professional technical knowledge and interdisciplinary insights, coupled with innovative thinking and digital literacy—has become paramount in driving value creation.

The artificial intelligence industry, teeming with infinite possibilities, exemplifies this need. Professionals in this field must not only be adept in machine learning, deep learning algorithms, and other technical domains but also understand industry application scenarios and possess interdisciplinary thinking abilities. They must be able to seamlessly integrate artificial intelligence technology with the needs of sectors such as healthcare, finance, and transportation. For example, in intelligent healthcare, talents should not only master advanced image recognition algorithms to assist in medical image diagnosis but also understand medical knowledge and how to collaborate effectively with clinicians. Digital literacy further enables them to adeptly use various data processing tools and cloud computing platforms, facilitating the efficient development and deployment of artificial intelligence solutions.

In the burgeoning digital creative industry, compound talents also shine brightly. Animation and game producers, for instance, must not only possess exquisite painting and animation production skills but also understand cultural history, psychology, marketing, and other multidisciplinary knowledge. They draw upon diverse cultural elements, use psychological principles to design compelling plots and characters, and then market their works to meet the growing spiritual and cultural needs of the populace, thereby creating substantial cultural industry value.

To cultivate such versatile talents, the education system is undergoing constant innovation. Universities are offering interdisciplinary courses and encouraging students to participate in industry-academia-research collaboration projects and innovation and entrepreneurship activities, thereby enhancing their comprehensive literacy and practical abilities. Enterprises, too, are intensifying their internal training programs to provide employees with continuous learning opportunities, enabling them to keep pace with technological innovation and continuously enhance their value creation capabilities.

In conclusion, the interweaving and collaboration of new productive forces and value production factors are opening a new chapter in value creation for human society. The productivity leap driven by digital technology has expanded the boundaries of human development, while the rise of emerging value production factors such as data, knowledge and technology innovation, and new talents has reshaped the underlying logic of value creation, paving the way for a future rich in potential and progress.

9.2 New Rules of Human Value

In the era of Value-Based Civilization, the new rules of human value emerge as pivotal forces, wielding a profound influence across diverse domains. These rules are not merely superficial adjustments but are deeply integrated into the very fabric of economic, cultural, social, and other spheres of human activity. Their integration is transformative, effectively reshaping the modes of human behavior and interaction in ways that are both subtle and significant.

Economically, these value-based rules foster a environment where ethical considerations and long-term sustainability take precedence over short-term gains. They encourage businesses to adopt practices that are not only profitable but also socially responsible, leading to a more balanced and equitable economic landscape.

In the cultural realm, the new rules promote a celebration of diversity and inclusivity, encouraging the exchange of ideas and the appreciation of different perspectives. This cultural shift fosters a more harmonious and understanding society, where mutual respect and cooperation thrive.

Socially, the impact of these rules is equally profound. They guide individuals and communities towards more compassionate and empathetic interactions, strengthening social bonds and fostering a sense of collective responsibility. This shift in social dynamics helps to create a more cohesive and supportive society, where individuals feel valued and connected.

Moreover, the new rules of human value act as a catalyst for societal progress, helping to navigate the complexities of modern life and address the challenges of our time. They provide a framework for decision-making that is grounded in ethical considerations and a long-term vision, enabling society to move forward smoothly and sustainably.

In essence, the new rules of human value are not just guidelines; they are the architects of a new societal paradigm. They are instrumental in building a world where economic prosperity, cultural richness, and social harmony coexist, driving humanity towards a future that is not only prosperous but also meaningful and fulfilling.

9.2.1 Integrity and Transparency: The Cornerstone of the Digital Economy

In the contemporary landscape, the digital economy is experiencing exponential growth, with the scale of online transactions and virtual asset exchanges expanding at an unprecedented rate. This surge has cemented these activities as integral components of the global economic matrix. However, the traditional trust-building model, reliant on face-to-face interactions and paper-based contracts, is increasingly inadequate in this digital realm. Consequently, integrity and transparency have ascended to the forefront as paramount factors ensuring the robust and healthy development of the digital economy.

Enter blockchain technology, a revolutionary solution to the trust conundrum in digital transactions. A case in point is the world of digital currencies, where every transaction involving Bitcoin is meticulously recorded on a distributed ledger. This ledger, accessible via blockchain browsers, offers an unparalleled level of transparency: anyone can trace the flow of funds, view transaction timestamps, and access a myriad of other pertinent information. Moreover, these records are immutable, effectively thwarting attempts at fraud, money laundering, and other illicit activities. The transparency and permanence of the record system make it exceedingly difficult for malicious actors to manipulate transactions, as any attempt to tamper with the records is instantly flagged by the system, rendering such efforts futile.

The impact of blockchain extends beyond digital currencies and into the realm of enterprise operations. Listed companies are now leveraging blockchain ledger technology to provide real-time, accurate financial disclosures to investors. This marks a significant departure from the past, where investors relied on periodic financial statements that often suffered from lag and the potential for data manipulation. The blockchain-based information disclosure system has revolutionized this landscape, ensuring that financial data is transparent and credible from its inception. Every fluctuation in income and expenditure, as well as changes in assets and liabilities, is immediately synchronized with investors. This real-time transparency enables investors to make informed decisions, facilitating the rational allocation of capital to efficient and reputable enterprises. In turn, this fosters a environment of healthy competition within the industry, driving progress and innovation.

In essence, blockchain technology is not merely a technological advancement; it is a cornerstone of trust in the digital economy. By enhancing integrity and transparency, it is paving the way for a more secure, efficient, and equitable economic future.

9.2.2 Diversity and Inclusiveness: The Engine of Cultural Integration

Globalization and the widespread adoption of digital technology have brought people around the world closer together, leading to frequent interactions among different cultures, races, and beliefs. In this context, the values of diversity and inclusiveness have become paramount, strongly promoting the integration of cultural innovation and social harmony.

The cultural creation sector exemplifies the power of these values. In the animation industry, for instance, Chinese animation has benefited from the experiences of many countries, incorporating complex human emotions and innovative storytelling techniques. This cross-pollination of ideas is not limited to animation; it is a trend seen across various creative fields.

Transnational game development teams, comprising members from diverse global backgrounds, are becoming increasingly common. A popular large-multiplayer online role-playing game, for example, blends various cultural elements in its game scenes, such as Chinese Xianxia landscapes, European medieval castles, and African grassland customs. Players can immerse themselves in a virtual world that celebrates cultural diversity.

The game's plot and tasks also draw from a rich tapestry of global stories, including Nordic mythology and the ancient Chinese Silk Road. Through gameplay, players naturally come into contact with and gain an understanding of different cultures, breaking down cultural barriers and facilitating global cultural interaction in an entertaining way.

This digital era, characterized by globalization and technological advancement, is fostering an environment where diversity and inclusiveness are not just valued but are actively shaping the future of cultural innovation and social harmony. As we move forward, these principles will continue to play a crucial role in creating a world where cultural exchange and mutual understanding are at the forefront.

9.2.3 Sustainable Development: The Future of the Earth

In the face of resource shortage and ecological deterioration, the concept of sustainable development has been integrated into all aspects of human production and life. Whether it is personal consumption, enterprise decision-making or national planning, it is taken as an important guide and carries the responsibility of human beings for the future of the earth.

As the main force of the economy, enterprises actively practice sustainable development. In the product design stage, in order to reduce pollution and cater to consumers environmental preferences, electronic equipment manufacturers abandon materials containing mercury, cadmium and other harmful substances, and choose bamboo and recycled plastics to make mobile phone shells and computer bodies. Such materials are not only environmentally friendly, but also give the product a unique texture. In the production process, automobile manufacturing enterprises optimize the production line by introducing intelligent manufacturing technology, reduce energy consumption, and vigorously invest in the research and development of new energy vehicles, gradually replace fuel vehicles with electric vehicles, reduce carbon emissions from the source, and cope with climate change.

Consumers awareness of green consumption is also on the rise and becoming an important force to promote sustainable development. When buying home appliances, consumers will give priority to energy efficiency labels, and energy-saving refrigerators and air conditioners are more popular; daily travel, shared bikes and new energy vehicles are popular, and green travel has become the norm. This change of consumption choice encourages enterprises to accelerate the green transformation and form a joint force for the whole society to build sustainable development in the future.

The construction industry is a key practice area for sustainable development. Green building standards comprehensively regulate the whole process of building from planning to operation. Modern green buildings use a large number of solar photovoltaic panels, using natural light and natural ventilation, reduce the dependence on lighting and air conditioning systems, and achieve energy self-sufficiency. Geothermal energy, biomass energy and other clean energy sources are also widely used in building heating and cooling, reducing energy consumption and helping urban low-carbon.

9.2.4 Reflections on Value-Based Civilization: Painting a Grand Blueprint for Sustainable Development

In an era where resources are increasingly scarce and ecosystems are gradually losing balance, the concept of sustainable development emerges like a ray of light, piercing through the fog to illuminate the path forward for humanity. This concept is no longer merely a slogan or a tagline; it has deeply rooted itself in every level of personal consumption, corporate decision-making, and even national planning, serving as a beacon of values guiding us towards the future.

Corporations, as the core drivers of economic activity, are undergoing a profound transformation. They are no longer solely focused on profit maximization; instead, they are integrating social responsibility and environmental sustainability into their core business strategies. This shift is not just a response to regulatory pressures or market demands but a genuine awakening to the importance of long-term, sustainable growth.

Consumers, too, are undergoing a significant change. They are becoming more conscious of their purchasing decisions, awakening to the consciousness of green consumption. They are beginning to pay more attention to the environmental attributes of products, tending to choose companies and products that embody a sense of social responsibility. This shift in consumer mindset is like a quiet revolution, subtly altering the landscape of the market and promoting the spread and practice of value-based civilization in a broader sphere.

The journey towards sustainable development is not without its challenges. It requires a concerted effort from all stakeholders—governments, businesses, and consumers alike. But the rewards are immense. A value-based civilization that prioritizes sustainable development is not only economically viable but also morally and socially enriched. It is a civilization that respects the planet, cares for its people, and fosters a sense of shared responsibility for the future.

As we move forward, let us remember that sustainable development is not just a goal to be achieved; it is a way of life to be embraced. It is a call to action for each and every one of us to play our part in painting this grand blueprint for a better, more sustainable world.

9.2.5 *Responsibility and Responsibility: The Guardian of Social Public Order*

As we enjoy the convenience and opportunities brought by digital technology, the sense of responsibility and accountability among enterprises and individuals becomes indispensable. This responsibility is crucial in protecting social public order, upholding good customs, and ensuring the steady development of society driven by technological advancements.

Technology giants, leading the charge in innovation, must also assume significant social responsibilities. For instance, while face recognition technology is widely applied across various fields, it brings along risks such as privacy invasion and discrimination. To mitigate these risks, enterprises must establish stringent ethical guidelines and regulatory frameworks, ensuring that technological progress aligns with societal values and norms. This approach not only fosters technological innovation but also maintains the public order and good customs of both the real and virtual worlds.

In the era of value-based civilization, the four new rules of human value—integrity and transparency, inclusiveness, sustainable development, and responsibility and accountability—are deeply interconnected and mutually reinforcing. These principles provide a clear and feasible path for the development of human society, guiding us through the complexities of the digital economy, cultural exchanges, ecological protection, and social governance.

By embracing these values, we can navigate the challenges of our time and move towards a future that is not only technologically advanced but also ethically sound and socially responsible. This is the essence of Value-Based Civilization, where the pursuit of progress is harmoniously balanced with the preservation of our shared human values.

9.3 A New System of Social Governance

With the robust advancement of Value-Based Civilization, the imperative to construct a new, corresponding social governance system has emerged as a pressing mission entrusted by the era. This envisioned system resembles a highly sophisticated, intricate, and efficient supermachine, designed to navigate the complexities of modern society. Leveraging cutting-edge digital technology, it seamlessly integrates the strengths of multiple stakeholders, fostering a collaborative approach to governance. Its primary objective is to achieve comprehensive and refined social management, adeptly addressing the increasingly complex and ever-changing societal needs. By doing so, it aims to ensure the harmony, stability, and sustained progress of society, creating a foundation for a more equitable and prosperous future. This transformative system is not merely a response to contemporary challenges but a proactive step towards crafting a governance model that is resilient, adaptive, and forward-looking, capable of steering society through the myriad challenges of the twenty-first century and beyond.

9.3.1 Digital Technology: A Tool for Accurate Decision Making and Efficient Execution

The profound integration of digital technology into the fabric of society has unlocked unprecedented avenues for the reform of social governance, emerging as a cornerstone for precise decision-making and efficient execution.

Acting as the vigilant "guardian" of urban operations, the city management department is harnessing the formidable power of big data analysis to gain a holistic and nuanced understanding of the city's every "pulse." By delving into vast troves of data—including real-time urban traffic flow fluctuations, dynamic variations in residential water and electricity usage, and precise environmental quality monitoring—the city's managers are bestowed with a "god-like perspective." This enables them to anticipate potential urban operational issues with remarkable accuracy and craft highly targeted policies and measures in response.

A prime example of this digital revolution is the intelligent transportation system, whose essence lies in the real-time collection and sophisticated analysis of traffic data. Sensors embedded throughout the urban road network, akin to sensitive nerve endings, continuously feed information on vehicle speed, congestion levels, traffic flow, and intersection dynamics to the data processing center. Leveraging these real-time insights, intelligent algorithms swiftly intervene, dynamically adjusting signal light durations to ensure smooth traffic flow on major roads, significantly reducing intersection wait times, and effectively alleviating urban congestion. Moreover, through the meticulous analysis of historical traffic data, the system can accurately forecast peak traffic hours and problematic road sections, enabling proactive traffic management, optimized routing strategies, and the maintenance of unimpeded urban traffic circulation.

In the realm of environmental governance, digital technology's role is equally pivotal. Environmental monitoring departments deploy high-precision sensor networks to conduct round-the-clock surveillance of air, water quality, soil, and other critical environmental elements. Each data point serves as a "barometer" of environmental health, offering real-time insights into the state of the environment. Upon detecting anomalies, such as an exceedance of air quality indices or a sudden spike in pollutant concentrations in a river, the system can rapidly and accurately pinpoint the source of pollution and promptly disseminate relevant information to environmental law enforcement agencies. This precise targeting enables swift action, prompting polluting enterprises to undertake timely rectifications, addressing environmental pollution at its root, and thereby enhancing the city's livability and ecological integrity.

9.3.2 Multi-Subject Coordination: A New Pattern of Joint Construction, Joint Governance and Shared Benefits

Social governance in the new era has long since bid farewell to the traditional model of single leadership led by the government, and turned to a new pattern of active participation and coordinated governance of multiple subjects such as social organizations, enterprises and the public.

In response to global challenges such as the public health crisis, the strong synergy of multiple actors is fully demonstrated. With its strong resource coordination and coordination ability, the government has quickly assumed the core leadership responsibilities such as prevention and control policy formulation, medical resource allocation, and social order maintenance. Medical institutions give full play to their professional advantages and undertake key tasks such as patient treatment, vaccine research and development, and medical supplies support. Medical device manufacturers are producing masks, protective clothing, testing reagents and other materials to ensure the supply of anti-epidemic materials; science and technology enterprises actively use big data and artificial intelligence to help epidemic prevention and control, develop epidemic monitoring platforms and intelligent diagnosis and auxiliary systems to provide scientific and technological support for epidemic prevention and control; logistics enterprises are around the clock to ensure that epidemic prevention materials and living materials can be quickly and accurately delivered to every place.

Community organizations and volunteers, as the backbone of the grassroots response, go to every corner of the community to assist in basic work such as personnel screening, temperature detection and publicity of epidemic prevention knowledge. They go door to door to register residents information to ensure that there is no dead end in the investigation of the community, measure the temperature of residents, check the health codes, build the first line of defense of epidemic prevention and control in the community, popularize epidemic prevention knowledge to residents through community radio and WeChat group, guide scientific epidemic prevention, and eliminate panic. It is this multi-subject collaborative governance model with the concerted efforts of the whole society that enables mankind to quickly form a strong response force in the face of the public health crisis, effectively contain the spread of the epidemic, and ensure social stability and tranquility.

9.3.3 System Construction Keeps Pace with the Times: Standardizing Behavior and Ensuring the Cornerstone of Development

To adapt to the rapid pace of development of the digital age, the institutional construction of social governance must keep pace with The Times and provide a solid guarantee for the orderly development of social governance.

The continuous improvement of cyberspace legislation has become a key link in regulating the behavior of the virtual world. With the popularity of the Internet, network infringement, data privacy leakage and other problems have become increasingly prominent, bringing many challenges to social governance. In order to deal with these problems, countries have accelerated the process of network legislation, and clarified the legal boundary and responsibility of network behavior. For example, for network infringement, the law specifies the standards of infringement, compensation liability, so that the victims in the network infringement can law; in the protection of data privacy, legislation strictly regulate enterprise and individual data collection, storage, use and sharing behavior, data processors must take necessary security measures to protect the user data privacy, once data leakage, the responsible person will be investigated for legal responsibility, to protect the legitimate rights and interests of citizens in cyberspace.

The vigorous development of industry self-discipline norms is also an important part of the construction of social governance system. As an important participant in the digital era, Internet enterprises are well aware of their social responsibilities, and they have joined forces to formulate industry self-discipline norms and strengthen self-restraint. In the field of content audit, major Internet platforms jointly formulate content audit standards, explicitly prohibit the spread of violence, pornography, false information and other harmful content, and ensure the healthy and positive content of the platform through the combination of artificial intelligence algorithms and manual audit. In terms of data use, enterprises have reached a consensus to prevent data abuse and maintain the healthy and orderly development of the data market.

The wide application of credit evaluation system provides an innovative and effective incentive and restraint mechanism for social governance. Whether individuals or enterprises, the credit performance in economic and social activities is included in the credit evaluation system, and the credit rating is assessed according to the credit score. People with good credit can enjoy preferential policies and convenient services in financial loans, administrative approval, market access, such as small and micro enterprises can obtain low-interest loans with good credit record, and grow rapidly; while trust-breakers face joint punishment, restricted in travel, consumption, investment, such as trust-breaking people are restricted from high consumption and high-speed rail. This credit mechanism of "encouraging for keeping trust and punishing for breaking trust" promotes the whole society to form a good fashion of cherishing credit assets and consciously abiding by social rules, and effectively promotes the improvement of social governance level.

9.3.4 Challenges and Response: The Breakthrough Road in Forging Ahead

Although the new system of social governance has made remarkable achievements in many fields, it still faces a series of severe challenges on the way forward. We need to take precautions and actively respond, so that we can achieve breakthroughs in forging ahead.

First of all, the problem of data island has become a major bottleneck restricting digital technology to enable social governance. As the data between different departments and industries are often independent, the lack of effective sharing mechanism, which leads to the scattered storage of massive data and fails to form a resultant force. If traffic data, environmental data and population data in urban management cannot be interconnected, it will be difficult to give full play to the maximum effectiveness of big data analysis and affect the formulation of accurate decision-making. In order to break the data island, the government should play a leading role, establish a unified data standard and sharing platform, and promote the orderly flow of data across departments and industries, so that data can be truly "alive" in social governance.

Secondly, the problem of unclear rights and responsibilities in the process of multi-coordinated governance of multiple subjects needs to be solved urgently. In practice, the boundaries and responsibilities of social organizations, enterprises and the public in social governance are sometimes not clear enough, and they are prone to buck-passing and repeated work, which reduce the efficiency of governance. To this end, it is necessary to further refine the rights and obligations of each governance subject through legislation and formulation of rules and regulations, and clarify the division of responsibilities in different governance fields and links, so as to ensure the orderly progress of coordinated governance.

Moreover, the technical problems and the complex and changeable network environment of cyberspace governance bring great challenges to the system construction. With the rapid development of network technology, such as 5G, block chain, artificial intelligence and other new technologies are emerging, the means of network crime are also increasingly updated, the traditional means of network governance is difficult to deal with. This requires the regulatory authorities to continue to strengthen technology research, keep up with the pace of technology development, innovate regulatory methods, use new technologies to improve cyber governance capabilities, and at the same time strengthen international cooperation to jointly deal with global cyber governance problems such as cross-border cyber crime.

To sum up, the new system of social governance has made solid steps in the empowerment of digital technology, multi-subject coordination and system construction, but it still needs to face challenges and continuously optimize and improve them. Only in this way can we adapt to the development needs of the era of Value-Based Civilization and provide a solid guarantee for the harmonious, stable and prosperous development of human society.

9.4 New Mode of Transaction Distribution

As the relentless tide of globalization surges forward, propelled by the rapid advancements in science and technology, the traditional trade distribution model finds itself confronting an unprecedented array of impacts and challenges. In this pivotal moment, a new trade distribution paradigm emerges, akin to a profound economic tempest that is sweeping across various industries with unparalleled force. This transformative storm is not merely reshaping the global economic landscape but is also infusing it with fresh vitality, thereby catalyzing economic development and ushering in a multitude of new opportunities and transformative changes.

The emergence of this new trade distribution pattern is characterized by its dynamic and innovative approach, which diverges significantly from the static and often inefficient methods of the past. It leverages cutting-edge technologies such as artificial intelligence, big data, and blockchain to streamline processes, enhance transparency, and reduce costs. This technological revolution is enabling businesses to operate with greater efficiency and flexibility, thereby adapting more effectively to the ever-changing demands of the global market.

Furthermore, this new paradigm is fostering a more inclusive and interconnected global economy. By breaking down traditional barriers and facilitating easier access to international markets, it is enabling small and medium-sized enterprises (SMEs) to participate more actively in global trade. This democratization of trade is not only creating new avenues for growth but is also promoting a more equitable distribution of economic benefits.

In addition to these tangible economic impacts, the new trade distribution pattern is also driving significant shifts in consumer behavior and preferences. As consumers become more informed and connected, they are demanding greater variety, quality, and sustainability in the products they purchase. This shift is prompting businesses to adopt more sustainable and ethical practices, thereby contributing to a more responsible and sustainable global economy.

Moreover, the new trade distribution pattern is fostering greater collaboration and cooperation among nations. By promoting dialogue and mutual understanding, it is helping to mitigate trade tensions and disputes, thereby creating a more stable and predictable global trading environment. This collaborative approach is essential for addressing the complex challenges of the twenty-first century, such as climate change, poverty, and inequality.

In conclusion, the new trade distribution pattern that is emerging in the context of advancing globalization and technological development is a game-changer for the global economy. It is not only reshaping the economic landscape but is also injecting new vitality, creating new opportunities, and driving transformative changes across various industries. As we navigate this new era, it is essential to embrace these changes and harness their potential for the betterment of humanity.

9.4.1 The Rise of the Digital Platform-Driven Trading Ecology

With the deep popularization of Internet technology, digital platform has become the core hub of modern transaction distribution, giving birth to a new transaction ecology. Take e-commerce giant Amazon, for example, which has built a huge and complex online trading empire, closely connecting suppliers and consumers around the world. On this platform, small and medium-sized enterprises have unprecedented market access opportunities, without having to build physical sales networks, just upload product information to Amazon platform to display products to billions of consumers around the world. A small manufacturer of handmade leather goods, in the United States, with its unique design and exquisite leather goods, sells its products to Europe, Asia and other regions through Amazon platform. Its annual sales have increased several times in just a few years, which not only revitalized the enterprise, but also created a large number of jobs.

At the same time, the digital platform uses big data and artificial intelligence technology to achieve accurate matching between supply and demand. By analyzing massive data such as users browsing history, purchase behavior and search keywords, Taobao, Jingdong and other e-commerce platforms accurately push product recommendation information for consumers in line with their interests and needs, greatly improving shopping efficiency and user experience. For merchants, this means that they can more accurately locate the target customer groups, optimize the product marketing strategies, and reduce the waste of marketing costs. For example, a beauty makeup brand in Tmall platform, with the help of the platform to provide data analysis tools, found its products in young women, especially in a second-tier cities consumers, so targeted to adjust advertising strategy, increase in social media, beauty makeup APP promotion, cooperate with beauty makeup bloggers live with goods, product sales rose rapidly, to significantly expand market share.

In the transaction payment link, the digital payment system has also brought about revolutionary changes. Third-party payment platforms such as Alipay and WeChat payment have been widely popularized in China, making mobile payment a standard part in Peoples Daily life. Consumers do not need to carry cash or bank cards, but can pay for shopping in a few seconds with just a smartphone. Whether it is breakfast consumption in street stores or luxury goods in large shopping malls, mobile payment provides a convenient and efficient solution. In the field of cross-border transactions, PayPal and other international payment tools also play a key role, breaking the bottleneck of cumbersome cross-border transfer procedures and high fees of traditional banks, enabling global consumers to complete cross-border shopping payment easily and quickly, and greatly promoting the prosperity and development of international trade. According to statistics, in China, the scale of mobile payment transactions has continued to grow rapidly in recent years, with an annual growth rate of more than 20%, which promotes the deep integration of online and offline consumption and becomes a new engine of economic growth.

9.4.2 Supply Chain Finance Optimizes the Transaction Capital Flow

Integral to the evolving landscape of transaction distribution, supply chain finance emerges as a beacon of innovation, offering strategic solutions to the perennial challenges of enterprise capital turnover and the optimization of industrial chain capital flow. In the traditional supply chain paradigm, enterprises, both upstream and downstream, often grapple with the dual specters of sluggish capital withdrawal and formidable financing hurdles. This predicament is particularly acute for small and medium-sized enterprises (SMEs), whose limited collateral and lower credit ratings frequently barricade their access to comprehensive bank loans, thereby imposing severe constraints on their production, operational efficacy, and growth trajectories.

In contemporary times, the supply chain finance model, anchored by core enterprises, is experiencing a renaissance. A case in point is the automotive industry, where large automobile manufacturers, as pivotal core entities, have forged robust collaborative alliances with a myriad of upstream parts suppliers and downstream dealers. Leveraging the credit endorsements of these core enterprises, banks and financial institutions are empowered to deliver a spectrum of financial services, including accounts receivable financing and inventory pledge financing, to the entire spectrum of supply chain participants.

SMEs, which often navigate the treacherous waters of extended payment cycles and capital turnover tribulations, find respite in the supply chain finance model. By capitalizing on the supply contracts inked with core enterprises, these SMEs can pledge their accounts receivable to banks, thereby securing swift financing. This influx of capital is instrumental in fueling raw material procurement, scaling up production, and ensuring the uninterrupted continuity of enterprise operations.

The advent of blockchain technology has further revolutionized the landscape of supply chain finance, enhancing its transparency and security. The immutable and traceable nature of blockchain's distributed ledger ensures that every transaction within the supply chain is verifiable and tamper-proof. This technological leap enables financial institutions to assess enterprise credit risk with greater precision, lower financing thresholds, and elevate financing efficiency.

Illustratively, in the agricultural products supply chain, blockchain technology records comprehensive information spanning from the planting source, picking, processing, transportation, storage, to the sales terminal. This end-to-end transparency allows financial institutions to gain a holistic understanding of the agricultural product lifecycle, facilitating more accurate and convenient financial support for agriculture-related enterprises. Such advancements not only bolster the stable operation of the agricultural products supply chain but also catalyze the healthy development of the agricultural industry, underscoring the transformative potential of supply chain finance in the modern economic ecosystem.

9.4.3 The Sharing Economy Expands the Transaction Boundaries

The sharing economy model, with its innovative ethos of resource sharing and collaborative consumption, has revolutionized the landscape of traditional transactions, carving out fresh business opportunities and expansive value spaces. A prime illustration of this paradigm shift can be seen in the realm of travel sharing, exemplified by Uber and Didi. These platforms have adeptly harnessed the potential of underutilized private vehicles, creating a seamless connection between private car owners and those in need of transportation. This synergy has given rise to a vast urban transportation network, where private car owners can supplement their income by offering rides during their spare time, while passengers benefit from enhanced convenience, a diversification of travel options, and competitive pricing that outstrips traditional taxis. This innovative model not only streamlines the allocation of urban transportation resources and alleviates traffic congestion but also fosters a new era of flexible employment. According to recent statistics, the global community of individuals earning part-time income through ride-sharing platforms has swelled to tens of millions, underscoring the model's widespread adoption and economic impact.

In the sphere of accommodation, Airbnb has spearheaded a residential revolution, transforming the way we think about lodging. By leveraging unused residential spaces—ranging from urban apartments to countryside villas and seaside cottages—Airbnb enables landlords to easily list their properties on a global platform, attracting tourists seeking unique and immersive experiences. This model offers travelers the chance to reside in environments that are more distinctive and authentic than traditional hotels, providing a taste of local life and the warmth of a "home away from home." The economic benefits for landlords are substantial, with some in popular European tourist destinations earning thousands of euros per month during peak seasons. Moreover, this cultural exchange fosters greater understanding and integration among diverse communities, as tourists engage with local lifestyles and customs, creating a symbiotic relationship between economic gain and cultural enrichment.

Beyond transportation and accommodation, the sharing economy has also given rise to shared office spaces, offering flexible and cost-effective solutions for entrepreneurs, freelancers, and small to medium-sized enterprises. Brands like WeWork have established a global network of shared offices, where businesses can lease individual workstations, offices, or conference rooms according to their needs. This model provides access to shared amenities such as reception areas, conference rooms, and leisure spaces, thereby lowering the barriers to entry and reducing operating costs. Such initiatives are catalysts for the thriving innovative and creative industries, driving economic diversification and marking a significant step forward in the evolution of the modern workplace. The sharing economy, in its various forms, continues to reshape industries and economies, highlighting the power of collaboration and resource optimization in the digital age.

9.4.4 Cross-Border E-Commerce Is Reshaping the International Trade Pattern

As a dynamic and burgeoning sector within the realm of international trade, cross-border e-commerce is rapidly redefining the global trade landscape, fostering an environment that is increasingly inclusive and accessible. Historically, small and medium-sized enterprises (SMEs) have encountered significant hurdles when attempting to engage in international trade, ranging from prohibitive trade costs and intricate customs procedures to the challenges of establishing reliable international sales channels. However, the advent of cross-border e-commerce platforms such as Kuomitong and Amazon's global stores has effectively dismantled these barriers, unlocking the vast potential of the international market for SMEs.

In China, numerous small and medium-sized manufacturing enterprises have leveraged these platforms to directly export high-quality products to consumers worldwide. A notable example is the Zhejiang Yiwu small commodity manufacturers, who have utilized the Speed Buy platform to distribute a diverse array of creative household products and accessories to regions including the Middle East, Europe, and Africa. These products, offered at a fraction of the price of local equivalents, have garnered immense popularity among overseas consumers, leading to a surge in exports. This growth has not only catalyzed the development of local industry clusters but also generated a substantial number of jobs, thereby fostering regional economic prosperity.

The advancement of cross-border logistics and distribution has further bolstered the growth of cross-border e-commerce. Logistics companies such as DHL and SF International have continuously optimized their cross-border logistics networks, incorporating advanced warehouse management systems and intelligent sorting technologies to enhance efficiency and reduce costs. For instance, a parcel sent from Shenzhen, China to Los Angeles, which previously took 15–20 days to deliver, can now be signed for in as little as 5–7 days through specialized logistics services, representing a significant improvement in timeliness and a reduction in logistics costs by approximately 30%.

Moreover, governments across the globe have recognized the potential of cross-border e-commerce and have implemented policies to encourage and support its development. These measures include simplifying customs declaration processes, establishing comprehensive pilot zones for cross-border e-commerce, and offering tax incentives, all of which have further fueled the sector's growth and prosperity. It is anticipated that in the coming years, the volume of cross-border e-commerce trade will continue to grow at a rapid pace, potentially accounting for more than one-third of total global trade volume. This trend positions cross-border e-commerce as a mainstream model of international trade, driving towards a more balanced and inclusive global economic development.

9.4.5 Challenges and Response: The Road to Forge Ahead

Although the new model of transaction distribution has shown great potential and vitality, it will inevitably face many severe challenges in its development process. It is urgent for all parties to work together to explore ways to solve them, so as to forge ahead in the tide of economic development.

First, digital platforms face data privacy and security issues. As digital platforms collect, store and process massive amounts of user data, the risk of data leakage has become increasingly prominent. In recent years, there have been data leaks of large digital platforms in many places around the world. For example, Facebook reported that billions of users information had been improperly obtained, causing strong public concerns about data privacy. On the one hand, it damages the rights and interests of consumers, leads to the abuse of personal information, and may face fraud, harassment and other problems; On the other hand, it also brings huge reputation loss and reduces user trust, which affects the business development of enterprises. To meet this challenge, digital platforms need to increase investment in technology and strengthen data security protection. At the same time, the government should improve relevant laws and regulations, strengthen supervision and law enforcement, severely punish the behavior of illegal collection and use of user data, make clear that the ownership of data belongs to users, and protect the legitimate rights and interests of users.

Secondly, supply chain finance has hidden dangers of credit risk transmission. In the supply chain finance system, the credit relationship between the upstream and downstream enterprises is closely linked. Once the risk of default occurs in a certain link, it is easy to be transmitted along the supply chain, causing systemic risks. For example, during the financial crisis in 2008, due to the collapse of the real estate market, a large number of construction enterprises closed down, and the suppliers providing raw materials could not recover the payment, which led to the credit contraction of banks to the upstream and downstream enterprises of the whole industrial chain. Many enterprises fell into financial difficulties and their production and operation were blocked. In order to prevent such risks, financial institutions need to establish a more perfect risk early warning mechanism, strengthen the credit evaluation of enterprises in all links of the supply chain, not only pay attention to the credit status of core enterprises, but also carry out dynamic monitoring of upstream and downstream small and medium-sized enterprises. At the same time, enterprises are encouraged to use financial derivatives to hedge risks, such as credit default swaps and other tools, to spread credit risks and ensure the sound operation of the supply chain financial system.

Moreover, the sharing economy is facing regulatory difficulties. As an emerging business model, the innovation and flexibility of the sharing economy make it difficult to adapt to the traditional regulatory model. For example, in some cities, the qualification audit and vehicle access standards of online ride-hailing drivers have caused disputes, and some drivers have not passed strict background checks, causing safety risks; the insurance and annual inspection of vehicles also lack clear norms, which can easily lead to traffic accident disputes. In addition, shared accommodation is faced with a lack of supervision in health and fire protection, which brings potential risks to consumers. Government departments should keep pace with The Times, innovate regulatory methods, and establish a regulatory framework adapted to the characteristics of the sharing economy. Strengthen inter-departmental coordinated supervision, such as joint law enforcement by transportation, public security, industry and commerce, health and other departments, formulate unified industry standards and norms, it not only protects the safety and rights and interests of consumers, but also creates a good policy environment for the healthy development of the sharing economy.

Finally, cross-border e-commerce faces trade policy uncertainty. In recent years, global trade protectionism has been on the rise and the trade policies of various countries have brought many uncertainties to cross-border e-commerce. The trade war initiated by the United States has imposed tariffs on many Chinese goods exported to the United States. The product costs of many cross-border e-commerce enterprises have risen sharply, their profit margins are compressed, and some enterprises are even forced to withdraw from the US market. At the same time, different countries have different regulations on tax policies, product quality standards, intellectual property protection and other aspects, which also increases the compliance cost of cross-border e-commerce enterprises. Cross-border e-commerce enterprises should pay close attention to the international situation and the policy trends of various countries, strengthen communication with industry associations and government departments, optimize supply chain management and reduce policy risks through overseas warehouse layout, local registered companies and other ways. Intergovernments should strengthen dialogue and consultation, coordinate trade policies through bilateral or multilateral trade agreements, reduce trade barriers, and create a stable and predictable development environment for cross-border e-commerce.

The new model of transaction distribution has opened a new chapter in economic development. Digital platforms, supply chain finance, sharing economy, cross-border e-commerce and other fields are booming, injecting strong impetus into the global economy. Despite these challenges, as long as all parties respond in the spirit of openness, innovation and cooperation, they will overcome many obstacles and push the economy into a more prosperous and sustainable future. In retrospect, in the early stage of mobile payment, all parties worked together to overcome security concerns and user habits, which made mobile payment rapidly promoted and became an indispensable part of people's lives. In the early stage of cross-border e-commerce, enterprises and the government worked together to solve logistics, customs clearance and other problems, which promoted the strong growth of cross-border e-commerce. These successful experiences have provided valuable lessons for meeting the current challenges and encouraged us to move forward, explore and innovate, and write a brilliant chapter of economic development.

9.5 A New Picture of Global Cooperation

Under the profound influence of value-based civilization rooted in shared principles of sustainable development and ethical governance, global cooperation is undergoing a far-reaching reshaping that fundamentally alters international relations architecture. This transformative wave manifests through concrete multilateral initiatives like the Paris Agreement's enhanced emission reduction mechanisms and the Global Minimum Tax framework adopted by 136 jurisdictions. A new picture of both opportunities and challenges is slowly unfolding, evidenced by the 78% surge in cross-border technology alliances since 2020 and simultaneous rise in trade protectionist measures across 62 economies. This paradoxical landscape presents unprecedented possibilities for knowledge-sharing while testing institutional capacities to balance national interests with collective welfare.

This wave of changes has covered many key areas, including politics, economy, culture, science and technology, creating complex interdependencies. In political spheres, the European Union's Carbon Border Adjustment Mechanism illustrates how environmental values reshape trade diplomacy, while ASEAN's consensus-building model demonstrates value-based conflict resolution. Economically, the explosive growth of blockchain-based Central Bank Digital Currencies (projected to reach $5 trillion in circulation by 2030) redefines monetary sovereignty concepts. Culturally, UNESCO's "Re|Shaping Policies for Creativity" program has enabled 48 nations to develop cultural economy blueprints aligned with SDGs. Technologically, the fusion of AI ethics frameworks (like the OECD's AI Principles adopted by 42 countries) with industrial innovation drives both progress and regulatory challenges.

Prompting countries to abandon the old isolated thinking mode, these developments have catalyzed remarkable policy shifts—exemplified by Japan's 2023 Strategic Collaboration Act that increased international R&D partnerships by 150%. Actively participating in the era of coordinated development, nations now engage in complex balancing acts: Germany's "Climate Partnerships" program simultaneously addresses emission reduction (targeting 65% by 2030) and technical capacity-building across 23 developing nations. Work together to respond to various global issues has produced innovative mechanisms like the Vaccine Alliance's dynamic pricing model, which improved low-income countries' access to malaria vaccines by 300% while maintaining pharmaceutical innovation incentives.

Strive to pursue the bright future outlined by a community with a shared future for mankind now materializes through concrete metrics. The Global Infrastructure Hub reports 214 transnational connectivity projects in development, potentially boosting collective GDP by $4.3 trillion annually by 2040. Simultaneously, the World Ethical Technology Initiative's global consultation framework has engaged 2.8 million citizens in shaping AI governance principles. These coordinated efforts, while facing challenges like the 34% increase in cross-border data flow disputes, demonstrate humanity's capacity to evolve collaborative paradigms that harmonize civilizational values with pragmatic problem-solving.

9.5.1 New Chapter of Economic Integration: Industrial Coordination and Resource Sharing

With the relentless march of globalization, the landscape of economic integration has undergone a transformative evolution, ushering in a new era where industrial coordination and resource sharing have emerged as the twin engines propelling the world economy towards unprecedented prosperity.

In the realm of manufacturing, the transnational industrial chain has reached a state of maturity, weaving an intricate global production network that is both robust and resilient. The automobile industry, a quintessential example of this global collaboration, showcases how the birth of a single car is a testament to the collective wisdom and resources of nations across the globe.

Germany, with its unparalleled expertise in engine manufacturing, infuses the vehicle with a powerful heart, ensuring robust performance and reliability. Japan, renowned for its precision electronic equipment technology, bestows the car with intelligent control systems, enhancing the driving experience with cutting-edge innovation. Meanwhile, China, leveraging its abundant labor resources and the vast, untapped potential of its consumer market, plays a pivotal role in reducing production costs and boosting efficiency.

Global automotive giants like Toyota and Volkswagen have masterfully harnessed the unique factor endowments of different regions, strategically establishing production bases worldwide to achieve a refined division of labor and seamless cooperation. Engines are meticulously researched and developed in Germany, where a deep industrial background and a pool of skilled talents guarantee exceptional performance. The electronic control systems are crafted in Japan, leveraging local advancements in electronics to ensure the vehicle remains at the forefront of intelligent innovation. The body assembly is undertaken in China, where relatively low labor costs and proximity to the consumer market contribute to cost-effectiveness and market accessibility.

This transnational industrial chain coordination model is a symphony of global collaboration, enabling automobile products to amalgamate the strengths of various countries, resulting in products of superior cost-performance that enjoy robust sales worldwide. This not only meets the diverse needs of consumers but also catalyzes the vigorous development of upstream and downstream industries, creating a plethora of employment opportunities and fostering economic growth.

In the realm of resources, the trend towards international sharing and cooperation has become increasingly pronounced. Oil, the lifeblood of the global economy, is extracted, transported, and sold through an intricate network involving numerous countries. The Middle East, with its vast oil reserves, stands as a major supplier to the world, having established long-term and stable cooperative relationships with nations across Asia, Europe, and the Americas.

Saudi Arabia, for instance, utilizes large tankers to deliver oil to every corner of the globe, providing indispensable energy support for the continuous operation of industrialized nations. In return, oil importers participate in infrastructure construction and oil exploration and development, creating a win-win scenario. European and American oil companies, for example, offer advanced exploration technology and equipment to Middle Eastern oil producers, aiding them in discovering new reserves, while securing stable oil supplies for themselves.

Moreover, the spirit of cooperation extends to the realm of renewable resources. Many European countries have joined forces to launch offshore wind power projects, sharing technological research and development results with nations like Denmark, the Netherlands, and Germany. This collaborative effort aims to overcome the technical challenges of offshore wind power, thereby promoting the growth of the clean energy industry and making significant contributions to the global fight against climate change. Researchers from various countries work hand-in-hand with enterprises, from the design and installation of wind turbines to their operation and maintenance. This collective wisdom has gradually reduced the cost of offshore wind power while steadily increasing its generation capacity, offering a viable pathway for the world's energy transformation.

In essence, the continuous advancement of globalization has not only turned a new page in economic integration but has also fostered an environment where industrial coordination and resource sharing are the cornerstones of global prosperity. As nations continue to collaborate and innovate, the future of the world economy looks brighter than ever.

9.5.2 Scientific and Technological Cooperation: Overcoming Problems and Sharing Achievements

In this modern era, characterized by the swift advancement of science and technology, global challenges have intensified, posing increasingly intricate scientific and technological difficulties that surpass the capabilities of individual nations. Amidst this backdrop, the surge in international scientific and technological cooperation has emerged as a pivotal支柱 in addressing these challenges and sharing the fruits of innovation.

Space exploration stands out as a quintessential example of this collaborative spirit. The International Space Station (ISS) epitomizes the remarkable achievements of global cooperation, with contributions from the United States, Russia, the European Space Agency, Japan, Canada, and other international and regional entities. These nations and organizations have invested substantial financial resources to assemble the world's top scientific research talents, culminating in the creation of this advanced space research platform. Each participant brings unique strengths to the table: the United States excels in the efficient transportation of personnel and materials; Russia, with its profound experience in manned space missions, ensures the long-term stability of the space station and provides state-of-the-art equipment for scientific experiments; Japan, for its part, is at the forefront of exploring emerging technologies like space robotics, contributing innovative insights. Astronauts aboard the ISS have conducted numerous cutting-edge scientific experiments across diverse fields such as life sciences, physics, and astronomy. Notably, in the realm of physics, experiments conducted under microgravity conditions are paving the way for the development of materials with entirely new properties. These groundbreaking research outcomes are generously shared, offering invaluable insights that fuel the expansion of our understanding of the universe and drive significant advancements in the human space industry.

In the realm of global public health, the importance of scientific and technological cooperation has been underscored by the COVID-19 pandemic. As the virus spread rapidly across the globe, the urgent need for vaccine development became apparent. Research teams worldwide overcame regional, racial, and political barriers, collaborating closely through online meetings and data sharing platforms. The Chinese scientific research team, for instance, responded swiftly, taking the lead in isolating the novel coronavirus strain and sharing its genetic sequence with the World Health Organization (WHO), thereby laying a solid foundation for global vaccine research and development. Developed countries like the United States, Germany, and the United Kingdom invested heavily in accelerating the vaccine clinical trial process, leveraging advanced experimental equipment and expertise to rapidly verify the safety and efficacy of vaccines. Meanwhile, developing countries such as India and Brazil contributed significantly by leveraging their rich clinical trial sample resources, thereby supporting a crucial aspect of vaccine development. This unprecedented level of collaboration among drug companies and research institutions led to the successful development of multiple COVID-19 vaccines in just a year. Through the WHO-led "COVID-19 Vaccine Global Access" (COVAX) initiative, these vaccines were distributed to low-income countries worldwide, ensuring a balanced supply of global disease resistance resources. This global effort not only highlighted the formidable power of scientific and technological cooperation but also demonstrated its vital role in safeguarding human health on a global scale.

9.5.3 Cultural Exchanges and Mutual Learning: Diverse Symbiosis and Inclusive Development

As a magnificent jewel in the crown of humanity's spiritual realm, cultural exchanges and mutual learning have flourished under the new paradigm of global cooperation. In this era, pluralism and inclusiveness resonate as the most powerful chorus, infusing a continuous flow of intellectual and emotional nourishment into the advancement of human civilization.

The film industry emerges as a pivotal portal for cultural exchange, transcending geographical boundaries. Hollywood, with its dazzling array of production techniques and captivating storytelling, has captivated audiences worldwide. Beyond disseminating American cultural values, it has become a magnet for cinematic talent from every corner of the globe, infusing multicultural elements that enrich the tapestry of its creations. Chinese directors and actors, for instance, have woven Eastern threads into the Hollywood fabric, adding a distinct Oriental hue to its palette. Meanwhile, Bollywood, India's vibrant film industry, has enchanted audiences across South Asia, the Middle East, Africa, and beyond, offering a vivid showcase of Indian religious culture and folk traditions. In recent years, Chinese cinema, with masterpieces like "Crouching Tiger, Hidden Dragon" and "Hero," has introduced the world to Eastern philosophy and aesthetics, sparking a global fascination with Chinese traditional culture. These cinematic exchanges foster creative dialogue, from scriptwriting and shooting techniques to post-production, driving the holistic innovation and evolution of film art. As a result, audiences are treated to a global feast of diverse cultural styles, brought to life on the silver screen.

The literary landscape, too, is abuzz with the vibrancy of cultural exchanges and mutual learning. World literature traverses linguistic borders through translation, with the four great Chinese classics now available in numerous languages, enchanting Western readers. Nobel laureate Mo Yan's works, steeped in Chinese local culture, have garnered international acclaim, drawing global attention to contemporary Chinese literature. Japanese author Haruki Murakami's novels, a harmonious blend of Western literary styles and Japanese cultural nuances, have garnered a worldwide readership. Literary translators, the unsung heroes of cross-cultural communication, construct robust bridges that enable readers of diverse backgrounds to surmount language barriers, fostering spiritual resonance and catalyzing a global renaissance in literary creation. Writers from varied nationalities engage in frequent exchange initiatives, sharing their creative journeys, drawing mutual inspiration, and facilitating a synthesis of literary styles that reflects the rich tapestry of our global cultural heritage. This interconnected web of cultural exchange and mutual learning not only celebrates the diversity of human expression but also strengthens the bonds of understanding and empathy that unite us as a global community.

9.5.4 Political Cooperation and Linkage: Addressing Challenges and Maintaining Order

In the interconnected era of globalization, a myriad of global challenges transcend national borders, rendering them insurmountable by the efforts of any single nation. In response to this imperative, political coordination has emerged as a vital necessity, serving as a robust safeguard for nations to collectively confront challenges, uphold world order, and ensure stability.

In the realm of climate change, the Paris Agreement stands as a monumental testament to the power of international political cooperation. Leaders from around the globe convened in Paris, navigating through intricate negotiations to reach a historic consensus. They pledged a solemn commitment to unite in the battle to curb greenhouse gas emissions and collectively tackle the formidable challenge of global warming. The agreement delineates customized emission reduction targets and a comprehensive action roadmap for each nation, explicitly stating that developed countries bear the responsibility to provide financial and technical assistance. This support is crucial in empowering developing countries to enhance their capacity to mitigate and adapt to the impacts of climate change. The annual UN Climate Change Conference has evolved into a pivotal forum for nations to report on their progress, share experiences in climate response, and deliberate on subsequent actions. This collaborative effort is instrumental in accelerating the global transition to a green economy and in safeguarding the planet for the benefit of future generations.

The fight against terrorism is another domain where international political cooperation is indispensable. Terrorist activities pose a grave threat to global peace and security. In the wake of the September 11 attacks, which sent shockwaves around the world, the UN Security Council swiftly enacted a series of anti-terrorism resolutions. Countries across the globe have intensified efforts in intelligence sharing, joint law enforcement, and military cooperation. The United States and its allies have conducted counter-terrorism military operations in the Middle East, aimed at dismantling extremist terrorist organizations and eroding their capabilities. China, for its part, is actively engaged in international counter-terrorism efforts, employing education and training to eradicate the roots of terrorism in Xinjiang, thereby maintaining regional stability. By sharing its counter-terrorism expertise, including effective measures and strategies for social integration, China contributes invaluable wisdom to the global counter-terrorism endeavor. Together, nations are working in concert to forge a peaceful and stable world order, resilient in the face of terrorism and secure for all humanity.

9.5.5 Challenges and Response: The Road to Forge Ahead

Although the new vision of global cooperation shows many bright prospects, the road ahead is full of difficulties and faces a series of severe challenges. It is urgent for all countries to work together to explore solutions and embark on a new journey.

First, the rise of trade protectionism has created many obstacles to economic integration process. In order to protect their own industries and jobs, some developed countries frequently introduce "great kills" of trade barriers, imposing tariffs, implementing technical barriers and implementing trade restrictions. The trade frictions between China and the US provoked by the US in recent years have had a huge impact on bilateral trade and the stability of the global industrial and supply chains. Many multinational enterprises have suffered from rising costs, reduced orders, blocked upstream and downstream industrial chains, and slowed the pace of world economic recovery. Taking the electronics industry as an example, the United States imposed high tariffs on Chinese exports, which increased the export cost of relevant Chinese enterprises and severely compressed the profit space. At the same time, the procurement cost of American importers increased, so they had to reduce the order volume, and the upstream and downstream enterprises had excess capacity and capital turnover difficulties. In this regard, all countries should strengthen communication and consultation under the WTO framework, make full use of the multilateral negotiating mechanism to resolve trade disputes, actively advocate the concept of free trade, firmly oppose unilateralism, and make every effort to maintain an open and inclusive global economic order.

Secondly, the problem of technology blockade and data hegemony in science and technology competition are increasingly prominent. In order to maintain their leading position, some scientific and technological powers have imposed technology blockade on other countries, restricted the export of high-tech products and banned the transfer of technologies. The US chip ban on Chinese Huawei has caused difficulties in the supply of high-end chips and hinder the development of china's communication technology industry. At the same time, in the digital field, some countries monopolize data resources, make unfair data rules, and arbitrarily infringe the data sovereignty and privacy of other countries. Some Internet giants have taken advantage of their own platforms to excessively collect user data and use the data for business competition or political manipulation, which has aroused widespread concern in the international community. To get rid of this dilemma, countries need to strengthen coordination of science and technology policies, jointly establish fair and reasonable international rules for data governance, encourage open and cooperation in science and technology, jointly explore cutting-edge technologies, and promote balanced development of science and technology around the world. International science and technology cooperation funds can be set up to fund joint scientific research projects and promote technology sharing and talent exchange.

Moreover, misunderstandings and conflicts caused by cultural differences and differences occur from time to time, which restricts the deepening of cultural exchanges and mutual learning. Different religious beliefs, values and customs are prone to friction in the communication and collision, and even lead to cultural conflicts in extreme cases. Due to the lack of in-depth understanding of other cultures, some western film and television works and literary works are distorted and misunderstood, causing dissatisfaction among the people in relevant countries. In pursuit of commercial gimmicks, some films stereotype foreign cultures and hurt the feelings of local people. To solve this problem, it is necessary to strengthen cross-cultural education, incorporate different cultural knowledge into the education system, cultivate students global civic awareness from an early age, enhance mutual understanding, respect and tolerance of different cultural groups, and promote the continuous deepening of cultural exchanges in an equal and friendly atmosphere. We can also hold multi-cultural festivals and international cultural exchange camps so that people can experience the charm of different cultures.

Finally, international political cooperation is faced with a lack of trust and differences of interests. In the handling of some global issues, countries have a crisis of confidence due to their political system, development stage, geopolitics and other factors, and it is difficult to reach consensus on capital input, responsibility sharing and action coordination. On the issue of financial assistance on climate change, funding commitments from developed countries have been delayed, causing dissatisfaction among developing countries. Some developed countries are reluctant to fulfill their financial commitments due to domestic political pressure or economic interests, leaving developing countries unable to cope with climate change. Countries should strengthen dialogue and communication, seek the greatest common interests, take practical actions to enhance mutual trust and work together to address global challenges. A coordination group on international problem solutions can be established to regularly assess national actions and coordinate the interests of all parties to ensure that global issues are properly resolved.

A new picture of global cooperation has gradually unfolded. Economic integration, scientific and technological cooperation, cultural exchanges and political cooperation have developed vigorously, but it also faces many challenges. Only by all countries joining hands and forging ahead with an open mind, cooperation and innovative thinking can we overcome many obstacles, broaden the path of global cooperation and create a better future for mankind. For example, in the early stage of response to COVID-19, due to the shortage of medical supplies such as masks, countries rapidly expanded production capacity by sharing production technology and coordinating supply chain, etc. In the construction of the International Space Station, countries overcome technical difficulties and political differences, continue to promote scientific research, and accumulate valuable knowledge for human exploration of the universe. These successful cases will provide useful lessons for future global cooperation and encourage all countries to move forward.

9.6 New Mechanisms for Global Governance

In today's era, the process of globalization is like a fast and unstoppable train, closely connecting countries around the world, making them interdependent and deeply integrated in many fields such as economy, culture, science and technology. However, accompanied by is global problems increasingly complex, as the car window emerging thorns, from the financial turmoil, trade friction, to the environmental deterioration of climate change, biodiversity, to security threats of terrorism, regional conflicts, and network chaos breeding information leakage, rampant network crime, all to the stability and development of human society pose a serious challenge. In this complex and changeable background, the traditional global governance model is becoming more and more inadequate, just like an old sailboat, it is difficult to move forward smoothly in the turbulent waves, and effectively respond to the endless problems. In order to meet the development needs of The Times, a new mechanism for global governance arises at the historic moment. It is like a sophisticated compass for the reshaping of the world order. It runs through many key areas, such as economy, environment, security and Internet, guiding the direction, committed to reshaping the international order and opening up ways for the sustainable prosperity of human society.

9.6.1 Innovation in Economic Governance: Stabilizing Finance and Coordinating Trade

The global economy is like a turbulent sea, where unpredictable waves of geopolitical conflicts, energy supply chain fractures, and technological revolution impacts constantly test the navigation capabilities of economic vessels. The complex and volatile situation makes the field of economic governance have to open the road of innovation, etc. Stabilizing the financial system and coordinating international trade are like two solid lighthouses, whose beams penetrate the fog of economic uncertainty. For instance, the 2022 European energy crisis triggered by the Russia-Ukraine conflict revealed how energy security impacts currency stability, while the COVID-19 pandemic exposed vulnerabilities in global supply chain coordination. Emerging economies represented by Vietnam and India are reshaping the global industrial landscape through intelligent manufacturing and digital services, creating new coordination needs for international economic governance. The International Monetary Fund's Special Drawing Rights (SDR) allocation plan and the World Trade Organization's digital trade rules negotiations demonstrate institutional innovation attempts under the new economic paradigm.

The shadow of the international financial crisis still exists, like a long dark cloud stretching from the 2008 subprime mortgage crisis to the 2022 cryptocurrency market crash, always warning people of the existence of risks. To prevent a recurrence of the crisis, global financial regulation has continued to strengthen. The introduction of the Basel Accord is like a strict financial defense line, with Basel III requirements increasing Tier 1 capital adequacy ratio to 6% and introducing liquidity coverage ratios (LCR) and net stable funding ratios (NSFR). This regulatory framework has prompted banks like Deutsche Bank and Citigroup to enrich their capital buffer through multiple channels such as preferred stock issuance and profit retention, enhancing their ability to resist risks. Central banks around the world have also taken action to strengthen macro-prudential management, flexibly using counter-cyclical adjustment tools like China's "macro-prudential assessment system" (MPA) and the Federal Reserve's stress testing. When the economy is overheated, such as during the 2021 U.S. real estate frenzy where housing prices rose 19% year-over-year, the market shows irrational prosperity and the asset bubble breeds, the central bank decisively tightened liquidity through measures like raising reserve requirements and reverse repo rates to cool the overheated economy. When the economic downturn, like the 2020 global economic contraction of 3.1% caused by the pandemic, market confidence is frustrated and the risk of credit exhaustion, the central bank quickly injected funds through mechanisms like the European Central Bank's Pandemic Emergency Purchase Programme (PEPP) which allocated €1.85 trillion, acting like a shot in the arm into the weak economy to boost market confidence. For example, after the 2008 financial crisis, the US financial market fell into panic, with the TED spread (measure of credit risk) soaring to 4.65% points, credit freeze and corporate financing difficulties. The Fed's emergency quantitative easing policy, with massive bond purchases reaching $4.5 trillion and tons of money pouring into the market, stabilized the panic in financial markets and prevented credit depletion from triggering a deeper recession. At the same time, financial regulators will focus on shadow banking, financial derivatives and other areas with high risk, strengthening supervision over money market funds (reforming Rule 2a-7), regulating credit default swaps (mandating central clearing through Dodd-Frank Act), and enhancing disclosure requirements for structured products. The 2021 Archegos Capital Management $10 billion meltdown revealed regulatory gaps in total return swaps, prompting new rules on family office oversight.

9.6.2 Environmental Governance Coordination: Addressing Climate Change and Ecological Protection

In the face of the urgent crisis of global ecological environment deterioration, the earth is like a sick giant whose vital signs are showing multiple organ failure. The Amazon rainforest known as the "lungs of the earth" has seen 17% of its area disappear in the past 50 years, while coral reefs—the "tropical rainforests of the ocean"—have experienced 50% mortality due to ocean acidification. The ecosystem is in danger, as evidenced by the 68% decline in global wildlife populations since 1970 recorded by WWF's Living Planet Index. Coordinated environmental governance has become an inevitable choice, requiring the establishment of a "planetary health management system" similar to human body health management. Countries must work together to tackle climate change, protect biodiversity, and "feel the pulse" through satellite monitoring systems and biodiversity databases, then "apply the right medicine" through precision governance for the health of the earth.

The Paris Agreement establishes the core framework for the global response to climate change, just like formulating a "contract" for all countries to jointly protect the earth's climate. This legally binding international treaty, ratified by 195 countries, requires signatories to submit nationally determined contributions (NDCs) every 5 years. The agreement's "ratchet mechanism" ensures that emission reduction targets and action plans are progressively strengthened in accordance with the agreement. For instance, the initial NDCs would have led to 3 °C warming, but the updated commitments aim to limit warming to 2.4 °C, demonstrating the dynamic adjustment feature of this global governance framework.

The EU takes the lead in setting up an ambitious roadmap for carbon emission reduction, launching the "Fit for 55" legislative package in 2021. This comprehensive plan includes revising the Emissions Trading System (ETS) to cover maritime transport, establishing a Carbon Border Adjustment Mechanism (CBAM), and mandating that all new cars sold be zero-emission by 2035. Aiming to be carbon neutral by 2050, the EU has allocated 30% of its €1.8 trillion budget to climate-related projects, making great strides towards a green future through the Just Transition Fund that supports coal regions in energy transformation.

To achieve this goal, renewable energy and solar power are increasing year by year at an impressive rate. The EU's renewable energy share reached 22% in 2022, with solar photovoltaic capacity growing 25% year-on-year. In Europe, the coastline and fields are providing clean energy on roofs and open spaces through innovative solutions like Netherland's solar bicycle paths and Portugal's floating solar farms on reservoirs. Germany's "Agrophotovoltaic" projects combine agriculture with solar power generation, to capture every ray of sunlight and convert it into electricity while maintaining 80% of crop yields through optimized panel arrangements.

As an important member of the European Union, Germany has implemented the energy transition policy (Energiewende) that reduced coal's share in electricity generation from 40% in 2015 to 27% in 2022. The country phased out coal power generation through a €40 billion compensation plan for coal regions, while simultaneously constructing massive offshore wind farms in the North Sea. Although this process is full of challenges including grid infrastructure upgrades and energy storage technological breakthroughs, its firm determination has never wavered, evidenced by the 65% renewable energy target for 2030 enshrined in law.

At the same time, supporting the new energy vehicle industry has become a national strategy. Germany offers up to €9000 subsidies for electric car purchases, coupled with 10-year tax exemptions. The government has invested €6 billion in building one million charging points by 2030, creating a charging pile network where 75% of gas stations now offer fast charging. These policies have boosted electric vehicle sales to account for 31% of new car registrations in 2023, making the figures of new energy vehicles in the streets and lanes increasingly common. Munich's public transport system even features autonomous electric shuttles, becoming a beautiful scenery in the city that combines technological innovation with environmental sustainability.

Developing countries are not willing to be left behind in this green revolution. India has installed 72 GW of solar capacity through its International Solar Alliance initiative, while Brazil is regenerating 12 million hectares of Amazon rainforest by 2030. China has set the "dual-carbon" goal (peaking emissions by 2030 and carbon neutrality by 2060) to demonstrate its responsibility as a major country, backed by the world's largest carbon market covering 4.5 billion tons of emissions. The country's new energy vehicle production exceeded six million units in 2022, accounting for 60% of global output, while its wind and solar power installations reached 650 GW—equivalent to the total capacity of the next seven countries combined.

Accelerate the adjustment of the energy structure through a dual approach. On the one hand, China has reduced coal's share in primary energy consumption from 72.4% in 2005 to 56% in 2022 through "ultra-low emission" technological transformations covering 950 million kilowatts of coal power units. On the other hand, it has built nine wind and solar power bases in desert regions, each with 100 GW capacity, while pioneering the world's first fourth-generation nuclear power plant using molten salt reactors. This structural transformation is supported by a green financial system that issued $120 billion in green bonds in 2022, accounting for 40% of the global total.

9.6.3 Linkage of Security and Governance: Counter-Terrorism, Peacekeeping and Epidemic Prevention and Control

Security knows no borders, like a spreading fire, once out of control, will affect the world. The interconnected mechanism for global security governance has played a key role in key battlefields such as counter-terrorism, peacekeeping and public health prevention and control. It is like firefighters fighting in different fire lines to protect world peace and stability.

The fight against terrorism has lasted for many years, and the international community has formed a network of close cooperation, like a tight net, giving terrorists no place to hide. The United Nations has led the formulation of a counter-terrorism strategy, just like a wise military adviser, coordinating intelligence sharing and joint law enforcement operations among all countries, integrating resources of all parties and pooling global strength. The SCO has built a strong defense line against terrorism in Central Asia. The member states regularly hold joint anti-terrorism exercises. Soldiers of all countries fight side by side on the training grounds, exchange anti-terrorism experience and technology, jointly fight against the "three forces", protect regional peace and stability, and keep Central Asia away from the haze of terrorism. Countries also strengthen the anti-terrorism measures, the United States deployed advanced security equipment in airports, stations and other public places, every passenger luggage and personal inspection, no suspicious items; strengthen personnel background review, using big data, artificial intelligence technology to identify potential risk personnel; European countries improve the allocation of police force, street patrol police, enhance the ability to deal with emergencies, strengthen the network anti-terrorism monitoring, real-time monitoring of network dynamics, prevent the spread of terrorist ideas, terrorism in the bud. In addition to the traditional anti-terrorism means, the international community is also paying more and more attention to the ideological field of anti-terrorism, through education and cultural exchanges, spread the values of peace and inclusiveness, and resist the breeding of terrorist ideas from the root cause. For example, some international organizations have launched anti-terrorism education programs, enter schools and communities, popularize anti-terrorism knowledge to young people and the public, and help them establish a correct world view and value.

Peacekeeping operations are an important force in maintaining world peace and stability. The UN peacekeeping forces are rushing to conflict areas like ambassadors of peace. They monitor the ceasefire as the two sides of the conflict, protect civilians, protect the innocent people, and revitalize the devastated areas. In Africa, the UN Mission in South Sudan provides humanitarian assistance to local people, distributes food and medicine, provides food and clothing, builds temporary shelter, provides shelter for displaced people, and restores social order. In the Middle East, the UN temporary forces in Lebanon patrol the border, measure the boundaries of peace with firm steps, and creates conditions for regional peace, let the light of peace penetrate the haze of war. In recent years, with the diversification of forms of regional conflicts, the tasks of peacekeeping operations are also expanding. In addition to traditional military peacekeeping, non-military tasks such as assisting local governments in election supervision and promoting national reconciliation have also been added, so as to promote peace and stability in conflict areas more comprehensively.

The global pandemic of COVID-19 is also a major test of the global public health security governance. It is like a sudden storm, testing the response ability of countries around the world. The World Health Organization has urgently coordinated national resources, like the commander of the battlefield against the epidemic, issuing prevention and control guidelines, providing scientific strategies, organizing cooperation in vaccine research and development, and gathering global scientific research forces. Countries support each other in medical supplies and show the spirit of solidarity and mutual assistance. China donated masks, testing reagents, vaccines and other materials to many countries, and batches of relief materials cross mountain and ocean, carrying the responsibility; European countries jointly purchase medical equipment, integrate resources, share experience, communicate online treatment plans with medical staff, jointly respond to the impact of the epidemic, highlight the global force against the epidemic, fight the epidemic and protect the defense of human health. In the process of epidemic prevention and control, the weak links in the global public health system have also been exposed, such as the shortage of medical resources and the insufficient production capacity of epidemic prevention materials in some countries. To this end, the international community has begun to reflect on and strengthen the construction of the public health system, including increasing investment in medical and health infrastructure, training professional medical personnel, and establishing a global reserve mechanism for epidemic prevention materials, so as to improve the global ability to respond to public health emergencies.

9.6.4 Network Governance and Joint Construction: Standardize Order and Ensure Security

With the rapid development of the digital era, cyberspace has become a new frontier of global governance, just like an emerging "digital continent". This virtual territory, spanning 5.3 billion internet users worldwide, encompasses critical digital infrastructure including submarine fiber-optic cables spanning 1.4 million km, 8000 cloud computing data centers, and satellite constellations providing global coverage. Regulating the network order and ensuring the network security have become the top priority, just like building a solid "wall" and "law" for this new continent full of opportunities and challenges. The "Great Firewall" systems deployed by over 40 nations, next-generation quantum encryption technologies, and AI-powered cyber threat detection platforms represent modern equivalents of medieval castle defenses. Meanwhile, digital legislation is evolving rapidly—132 countries have enacted cybersecurity laws since 2015, with China's Cybersecurity Law and Singapore's Cybersecurity Act serving as model frameworks for balancing innovation and control.

The international community has actively formulated rules on cyberspace through multi-layered governance architectures. The UN's Group of Governmental Experts (GGE) has passed resolutions recognizing cyber sovereignty as an extension of national sovereignty, particularly emphasizing developing nations' rights to protect digital assets—a principle tested during the 2022 Ecuadorian financial system cyber siege where international law enabled data sovereignty recovery. The European Union's GDPR has created ripple effects beyond its borders through its "Brussels Effect", forcing 5000+ multinational corporations to redesign data architectures, exemplified by Microsoft's $20 billion global data center expansion for regional compliance. The United States Cyber Command's "Defend Forward" strategy has operationalized preemptive cyber defense, demonstrated by the 2023 neutralization of ransomware infrastructure targeting 50 Midwest hospitals. Cross-border collaboration mechanisms have matured significantly, with the China-US Cybersecurity Dialogue facilitating joint operations like "Project Shadowhunter" that dismantled APT41's cryptocurrency laundering networks across four jurisdictions. The Paris Call for Trust and Security in Cyberspace now endorsed by 1200 entities illustrates growing consensus, while technical standardization bodies like IETF and IEEE work to embed governance principles in protocol design through initiatives like RFC 8890 on digital sovereignty considerations.

The transnational nature of cyber crime has prompted countries to strengthen law enforcement cooperation through innovative institutional frameworks. INTERPOL's Global Cybercrime Program has established 24/7 operational hubs in Singapore and Bogotá, processing over 1.2 million threat indicators monthly through its I-24/7 secure network. The 2022 Operation HAECHI IV across 34 countries demonstrated new models of asset recovery, seizing $130 million in cryptocurrency from online fraud rings. Mutual Legal Assistance Treaties (MLATs) have been modernized through digital evidence sharing protocols like the CLOUD Act agreements, enabling Philippine authorities to access Meta servers in Ireland during the 2023 election interference investigations. Regional alliances are developing specialized capabilities—ASEAN's Cyber Capacity Development Program has trained 800 digital forensic experts, while Europol's Joint Cybercrime Action Taskforce (J-CAT) coordinates simultaneous takedowns across multiple jurisdictions. Private sector integration reaches new depths through programs like Amazon's Cybercrime Cloud Evidence Lockers, which preserve chain-of-custody for 140 law enforcement agencies. Emerging technologies bolster these efforts: blockchain analytics tools like Chainalysis Reactor helped trace North Korean Lazarus Group's $625 million crypto heist across 12 exchanges, while AI-powered linguistic analysis exposed 83% of dark web marketplace operators through stylometric fingerprinting.

9.6.5 Challenges and Response: The Way to the Thorns

The new global governance mechanism has encountered many obstacles on its way forward. It is urgent for countries to build consensus and work together to overcome difficulties to overcome difficulties and sail to the bright future.

First of all, the phenomenon of "free riding" in international cooperation occurs from time to time, which weakens the effectiveness of governance. In the field of tackling climate change, some countries try to enjoy the fruits of other countries emissions reduction, and they are reluctant to invest in capital and technology, leading to the slow progress of global emission reduction targets. In this regard, an incentive and restraint mechanism should be established to provide financial and technical assistance and rewards to the countries that actively implement the contract, impose sanctions on the countries that violate the rules, and strengthen the sense of responsibility sharing. A special international climate fund could be set up to allocate funds based on countries performance on emissions reduction, and to impose trade restrictions or carbon tax penalties on countries that fail to meet the standards to encourage real action.

Second, the great-power game brings uncertainty to global governance. The US pursues "America First" in international affairs and frequently withdraws from groups and breaks agreements, such as the Paris Agreement and the Iran Nuclear Agreement, which disturbs the international cooperation order and causes geopolitical unrest. Countries need to strengthen communication and coordination through multilateral dialogue platforms, guide the actions of major countries and use the vision of a community with a shared future for mankind, and resolve conflicts. For example, at the G20 summit, world leaders held in-depth discussions on global issues, sought consensus, coordinated actions, and provided constructive solutions for resolving international disputes.

Moreover, the development of emerging technologies brings new risks and tests the wisdom of governance. Cutting-edge technologies such as gene editing and artificial intelligence not only benefit mankind, but also cause ethical disputes and security risks. For example, gene editing and baby events impact the bottom line of ethics, and the weaponization of artificial intelligence threatens human security. The international community should make arrangements in advance, formulate forward-looking regulations and policies, guide the good development of technology, and establish a technology risk assessment and early warning mechanism to prevent potential crises. An international expert committee can be set up to conduct ethical review and risk assessment of emerging technologies, and release early warning information in a timely manner to provide reference for countries to formulate relevant policies.

Finally, there is unequal participation in the global governance system, and the voice of developing countries is limited. In the international financial institutions, the developed countries occupy the dominant position, and their decisions are biased towards their own interests. We need to improve the representation of developing countries, reform the governance framework, and make global governance more equitable and representative. We can also establish a capacity-building mechanism for developing countries to participate in global governance, and actively enhance the participation capacity and voice of developing countries in global governance through training and technical assistance.

The new mechanism for global governance has embarked on its journey and played a key role in various areas, but it faces many challenges. Only by all countries working together, upholding the principles of openness, inclusiveness and win-win cooperation, and improving governance mechanisms can we effectively respond to global issues and create a better future for mankind. In retrospect, in the Ebola epidemic prevention and control, all countries joined hands to support Africa and successfully contain the spread of the epidemic. Under the promotion of the International Telecommunication Union, the global 5G technology standards were gradually unified to promote the development of the communication industry. These successful practices have accumulated valuable experience for global governance and illuminated the road ahead.

9.7 Global Value Reconstruction

Under the continuous impact of the wave of globalization and the intensive force of global problems, the global value system is undergoing profound reconstruction through three-dimensional evolution. This transformative process manifests through three parallel dimensions: the vertical penetration of technological revolution (reshaping production relationships through AI and blockchain), the horizontal expansion of civilizational dialogue (evidenced by the Confucian Institute's global layout and European cultural protectionism), and the spiral upgrade of institutional frameworks (reflected in the Paris Agreement's iterative improvements). This change is not accidental, but the inevitable demands of time development, it is like a static depth of thought revolution, quietly penetrated into all aspects of human society. The silent metamorphosis manifests through three concentric circles of influence: the inner circle of individual cognitive frameworks (evolving from national identity to global citizenship), the intermediate institutional layer (witnessing WTO reform and BRICS New Development Bank establishment), and the outer civilizational dimension (emerging hybrid cultural forms like K-pop's global-local fusion). From the subtle transformation of ethical concepts, to review the cultural value, to the comprehensive innovation of social development concept, all reveal the human to cope with the complex world situation of profound reflection and actively explore. The ethical metamorphosis unfolds through paradoxical coexistence: bioethical committees simultaneously endorsing genetic editing while establishing red lines, corporate ESG standards balancing profit and sustainability. Cultural reevaluation demonstrates schizophrenic characteristics—the Louvre digitally preserves artifacts while fighting cultural appropriation lawsuits, and UNESCO protects intangible heritage as TikTok spawns new folk art forms. Development paradigm innovation exhibits hybrid vigor, seen in China's dual-carbon strategy merging ancient ecological wisdom with smart grids, and California's circular economy models combining indigenous knowledge with IoT solutions.

9.7.1 Ethical Innovation: Diversity and Responsibility Sharing

Traditional ethical concepts have undergone unprecedented refinement in the melting pot of globalization, and the new ethical code of multiple inclusiveness and responsibility sharing has gradually emerged and become a new beacon guiding human behavior. This transformation manifests profoundly in three dimensions: cross-cultural collision, institutional innovation, and value reconstruction. Like ancient silk roads being revitalized in the digital age, ethical systems rooted in agricultural civilizations now face multidimensional challenges from industrial, informational, and ecological civilizations. The 2021 UNESCO Global Ethics Report revealed that 78% of countries have introduced multicultural clauses in their legal systems over the past decade, demonstrating institutionalized ethical evolution. From the ethical debates surrounding AI development to the moral dilemmas in global vaccine distribution, humanity is collectively sketching a new coordinate system of civilization.

In the increasingly frequent cultural exchanges, people of different races, religions and regions frequently interact with the original single cultural frameworks. The 2015 European refugee crisis became a monumental ethical stress test: while Germany accepted over a million refugees, creating the "Wir schaffen das"(We can do this) model of social integration, Scandinavian countries implemented cultural mentorship programs pairing local families with newcomers. Contrastingly, Singapore's "racial harmony policy" established in 1960s serves as a preventive ethical framework, mandating ethnic ratio quotas in public housing. The ethical challenges of cross-cultural interaction are vividly reflected in corporate environments too. Tech giants like Google and Microsoft have developed "cultural intelligence" training modules, teaching employees to navigate differences in communication styles—from the high-context subtlety of Japanese business culture to the low-context directness of Dutch negotiations.

Some ethical judgments from a cultural perspective have become unsustainable. Diversified tolerance means abandoning prejudice and narrowness, and accepting values and lifestyles in various cultural backgrounds with an open-minded mind. The paradigm shift is epitomized in medical ethics: where Western individualism once dominated patient autonomy concepts, the 2020 pandemic saw East Asian "collectivist triage ethics" gain global recognition. Educational systems showcase similar evolution—Finland's 2016 curriculum reform incorporated "intercultural reasoning" modules, while Canada's Truth and Reconciliation Commission redesigned history textbooks to include indigenous perspectives. Culinary culture provides a metaphorical mirror: the UNESCO-listed "Mediterranean Diet" now integrates North African spices, Anatolian cooking techniques, and Iberian ingredients, creating a gastronomic ethics of coexistence.

For example, on the issue of immigration, many countries are gradually abandoning the attitude of rejection and recognizing the rich diversity brought by immigrants to the local society, whether it is the innovative inspiration in the field of culture and art, or the complementary vitality of the labor market. London's 2022 Cultural Capital Report showed immigrant communities contributed 38% of the city's new patent applications and 65% of Michelin-starred chefs. In Silicon Valley, 57% of tech startups have immigrant co-founders, driving what economists term "diversity dividends." Germany's dual vocational training system ingeniously integrates immigrant youth, pairing Syrian refugees with aging master craftsmen, simultaneously addressing skills shortages and cultural transmission. Singapore's "kampung spirit" revitalization project transforms immigrant neighborhoods into living museums of cross-cultural pollination.

In some European cities, by establishing multicultural communities, they provide space for immigrants to integrate, promote mutual understanding and cooperation among residents of different cultural backgrounds, and make food, music and art forms from all over the world blend together, blooming with colorful cultural flowers. Rotterdam's "Cultural Corridor" initiative transformed abandoned docks into 17 themed cultural plazas, each representing an immigrant group's heritage while hosting fusion festivals. Barcelona's "Superilles" (superblocks) urban redesign created hybrid spaces where Moroccan tea houses neighbor Catalan modernism galleries, featuring weekly "cultural potluck" events. The EU's INTERREG program funds cross-border cultural incubators, like the Danube Creative Basin connecting 10 countries' artisans through blockchain-enabled co-creation platforms. These experimental communities function as ethical petri dishes, cultivating new social grammars for coexistence.

The concept of responsibility-sharing is particularly critical in addressing global challenges. No country is immune from rising sea levels to the transnational spread of infectious diseases. The 2023 Global Risk Report identifies "climate inequality" as the century's defining ethical challenge: while Small Island Developing States (SIDS) contribute less than 1% of emissions, they face existential threats. This moral calculus drove the "Loss and Damage Fund" establishment at COP28, operationalizing climate ethics through $100 billion annual commitments. Similarly, the COVID-19 mRNA vaccine technology transfer hub in South Africa epitomizes pharmaceutical ethics globalization, increasing African vaccine production capacity from 1% to 15% within 2 years.

The international community has been deeply aware that all countries need to work together to respond and shoulder their responsibilities. The Kyoto Protocol and the Paris Agreement are vivid practices of sharing responsibility in the field of climate change. Countries commit to emission reduction targets according to their own development stages through innovative mechanisms like Brazil's Amazon Fund (financed by Norway and Germany) and China's South-South Climate Cooperation Fund. The 2025 Global Carbon Budget framework introduces "ethical emission quotas," combining historical responsibility with technological capability metrics. Beyond environmentalism, digital responsibility sharing emerges through the EU's AI Act and Singapore's Model AI Governance Framework, creating transnational ethical standards for algorithmic transparency. Even in space governance, the Artemis Accords establish protocols for lunar resource sharing, preventing cosmic colonialism through anticipatory ethics.

9.7.2 Remodeling of Cultural Values: Inheritance, Innovation, Exchange and Mutual Learning

In the spiritual map of human civilization, cultural values stand as an unshakable cornerstone at the precipice of epochal transformation. As waves of global value reconstruction sweep across the world, cultural value systems are embracing this metamorphosis with open arms. The core driving forces behind this profound transformation emanate from two eternal principles embedded in the genetic code of civilizations: the symbiotic dance between inheritance and innovation, and the dynamic interplay among diverse cultures.

On the dimension of civilizational inheritance, ethnic cultures are exploring new growth points between their traditional roots and modern soil. Take Chinese civilization as an example—this five-thousand-year-old cultural treasury is undergoing unprecedented creative transformation. The Palace Museum employs digital twin technology to reproduce the dawn and dusk of the Hall of Mental Cultivation in virtual reality spaces, allowing visitors to touch the temperature of history in digital corridors. The Dunhuang Academy's "Digital Patron" initiative leverages blockchain technology to enable public guardianship of cultural heritage, breathing new life into millennium-old murals in the digital realm. The Hanfu movement has demonstrated remarkable cultural reconstructive power, evolving from mere costume restoration into a complete cultural ecosystem encompassing ritual systems, traditional craftsmanship, and festival ceremonies. On Bilibili's Hanfu-themed live streams, young designers integrate quantum mechanics concepts into clothing patterns, creating the unique aesthetic of "cyber Hanfu."

At the forefront of cultural innovation, the fusion of technology and art is birthing unprecedented cultural forms. Japan's animation industry combines the planar aesthetics of ukiyo-e with CG technology, creating time-space interlaced visual spectacles in films like Your Name. The resulting "pilgrimage tourism" economy generates over $20 billion annually for Japan. South Korea's entertainment industry has built an "idol industrial system" that uses big data analysis to precisely capture global aesthetic trends. BTS employs holographic projection technology in their music videos to stage trans-dimensional performances on virtual stages, with their social media interactions surpassing even those of the United Nations official accounts.

In the dimension of cultural exchange, digital technologies are reshaping humanity's conversational modalities. The #TraditionalCultureChallenge on TikTok has amassed over 50 billion views, as Nigerian youth reinterpret Yoruba drumming with electronic music and Brazilian dancers integrate samba rhythms into Chinese kung fu, creating the unique "electro-martial arts" genre. The Venice Film Festival's VR competition unit enables Kenyan directors to reconstruct oral epics through virtual reality, breaking traditional documentary boundaries with immersive storytelling. At the Frankfurt Book Fair's digital copyright platform, Arabic magical realist novels and Latin American literary classics achieve instant translation and global distribution.

This deep integration is giving rise to new cultural paradigms: French designers apply Miao silver craftsmanship to high-end jewelry; Indian yoga combines with Nordic minimalism to form "digital meditation" trends. When Italian opera and Peking opera stage their first joint performance in the cloud, and African drumming resonates with electronic synths at music festivals, humanity is creating a new cultural ecosystem transcending single civilizational genes. This blend is not mere elemental overlap, but rather a cultural genetic mutation sparked through collision—much like the Dunhuang art born from ancient Silk Road exchanges, now blossoming even more brilliantly in the digital age.

Standing at this new nodal point of civilizational evolution, the reconstruction of cultural value systems concerns not just the continuation of traditions, but the construction of humanity's shared future. As civilizations freely converse in digital spaces and cultural innovation transcends geographical and historical boundaries, we are witnessing a new awakening of the human spirit. This awakening is not a simple replication of the past, but the creation of new possibilities through inheritance, and the gestation of a new chapter in civilization through blend. Just as the flying apsaras in Dunhuang murals remain stunning after millennia, the eternal charm of cultural values lies precisely in their ceaseless self-renewal and transcendence.

9.7.3 Change of the Concept of Social Development: Sustainable Development and Fairness and Justice

Amidst the pressing challenges of resource scarcity and escalating wealth disparities, the global narrative on social development is undergoing a profound transformation. No longer confined to the singular pursuit of economic growth, the discourse is expanding to embrace a more holistic approach that prioritizes sustainable development, fairness, and social justice. This shift is driven by a growing recognition of the need to address the deeper, structural issues that underpin social inequality and environmental degradation.

In response to these challenges, the concept of social development is moving forward, with a focus on creating resilient and inclusive communities. The International Labor Organization's Global Commission on Future of Work is at the forefront of this movement, emphasizing a "human-centered agenda" that promotes just transition mechanisms for workforce reskilling. This is exemplified by initiatives like the EU's Digital Europe Programme, which is investing 9.2 billion to support the development of digital skills and the transition to a more sustainable economy.

This holistic vision is anchored in the belief that true development transcends mere economic metrics, encompassing the broader spectrum of human experience—health, education, culture, and the environment. In this new era, governments, businesses, and civil society are converging towards a shared understanding: that the well-being of our planet and its inhabitants is intrinsically linked. This realization is driving innovative policies and practices that aim to balance economic growth with social and environmental sustainability.

The Paris Agreement and the Sustainable Development Goals (SDGs) reflect a global commitment to tackling climate change and preserving biodiversity, while also addressing issues of poverty, inequality, and social justice. These international frameworks provide a roadmap for countries to work together towards a common goal, fostering a sense of global solidarity and cooperation.

In essence, the journey towards a sustainable, equitable, and just future is a multifaceted one, requiring the collaboration of all sectors of society. It is a journey that acknowledges the interdependence of all life on Earth and seeks to build a world where prosperity is shared, and the dignity of every individual is upheld. This vision is not just a dream but a tangible goal within our reach, if we choose to act with wisdom, compassion, and unity. It represents a collective endeavor to reshape the very foundations of our societies, ensuring that progress benefits everyone, leaving no one behind. This transformative approach to social development is paving the way for a more sustainable, equitable, and just future for all.

9.7.4 Challenges and Responses: Forge the Road Ahead

The road to global value restructuring is fraught with multifaceted challenges that demand unprecedented international coordination. While the vision of establishing shared ethical frameworks across nations appears theoretically achievable, practical implementation faces formidable barriers rooted in historical legacies, geopolitical competition, and divergent development priorities. Economic disparities between Global North and South nations create asymmetric stakes in value system reforms, with developing countries often prioritizing poverty alleviation over abstract ethical debates. The digital revolution's uneven distribution further complicates matters, as technology giants' data monopolies clash with national sovereignty concerns. Recent trade wars over semiconductor supply chains and controversies surrounding cross-border data flows exemplify how commercial interests frequently override collaborative value-building efforts. Moreover, the resurgence of protectionist policies in major economies demonstrates how short-term national agendas can derail long-term global governance initiatives. Addressing these complex interdependencies requires not just diplomatic dialogue but structural reforms in international financial systems and technology governance frameworks.

First and foremost, systemic value conflicts manifest across multiple dimensions, creating persistent roadblocks in forging global consensus. Beyond the obvious religious-political divides, subtler cultural undercurrents significantly impact ethical perceptions. The Confucian emphasis on collective harmony versus Western individualism's focus on personal autonomy creates fundamentally different approaches to privacy rights and social responsibility. In technological ethics, these differences crystallize dramatically:

Gene editing controversies reveal deep philosophical rifts: The 2018 He Jiankui CRISPR baby scandal exposed conflicting regulatory paradigms, with China initially imposing a three-year research ban, while the EU has comprehensive genetic engineering directives; religious opposition from Catholic bioethicists contrasts with the technological optimism of transhumanist advocates; urgent health needs in developing countries (such as malaria resistance gene research) challenge the precautionary principle of developed nations. Euthanasia legislation showcases irreconcilable value systems—the Groningen Protocol for infant euthanasia in the Netherlands remains taboo in Confucian-influenced societies; the expansion of Medical Assistance in Dying (MAID) in Canada sparks constitutional challenges from disability rights groups; the universal prohibition by Islamic scholars contrasts with the increasingly accepting attitudes in secular European nations. New dilemmas in AI governance are also emerging—the EU's risk-based AI legislation conflicts with China's development approach emphasizing technological sovereignty; debates on algorithmic bias pit Western notions of individual fairness against Eastern models of collective welfare; indigenous data sovereignty movements challenge the data extraction practices of multinational corporations. To bridge these divides, a three-tiered multilateral mechanism should be established: Philosophical Arbitration Panels—combining the regional centers of the Elders' InterAction Council with UNESCO's

Bioethics Committee; Science Diplomacy Incubators—modeled on CERN's collaborative structure, focusing on frontier technology governance; Cultural Codebreaking Task Forces—teams of anthropologists and AI ethicists decoding value systems through computational social science. The Geneva-based Global Bioethics Forum has demonstrated partial success through its "Red Line/Blue Sky" framework, separating non-negotiable ethical boundaries from areas permitting cultural variation. However, lasting solutions require overhauling the current fragmented governance architecture, potentially through a reformed UN Trusteeship Council adapted for technological civilization challenges.

Secondly, the constraints of interest groups bring heavy resistance to the value reconstruction. This challenge manifests particularly in energy transition scenarios where entrenched economic forces create systemic barriers. For instance, in the United States, coal industry conglomerates invested $1.5 billion in political lobbying between 2010–2020, successfully delaying carbon emission reduction legislation for nearly a decade. The German automotive industry coalition, representing traditional combustion engine manufacturers, systematically undermined electric vehicle subsidy programs through strategic media campaigns exaggerating battery safety concerns. These protectionist behaviors often employ sophisticated tactics: fossil fuel giants establish "greenwashing" subsidiaries to feign environmental responsibility while continuing core extractive operations; energy lobbyists infiltrate regulatory committees through "revolving door" appointments; corporate-funded think tanks produce biased research questioning renewable energy reliability. A 2023 OECD report revealed that 68% of G20 nations experienced fossil fuel industry interference in their net-zero roadmaps. To counter this, Sweden's innovative "sunlight legislation" mandates real-time disclosure of all stakeholder meetings in energy policymaking, creating a publicly accessible digital ledger. Simultaneously, South Korea's "New Industry Incubation Fund" demonstrates effective government leadership by allocating $7.2 billion specifically to bypass traditional energy monopolies, directly supporting fusion energy startups and smart grid innovators.

Moreover, the imbalance of education popularization restricts the wide dissemination and internalization of global values. The education gap creates cognitive chasms that perpetuate value disparities—UNESCO data shows 263 million youth in sub-Saharan Africa lack access to secondary education, while in conflict-ridden regions like Yemen, 78% of schools remain non-functional. This deprivation cycle manifests starkly: pastoral communities in Mongolia's Gobi Desert remain unaware of climate change concepts due to outdated 1990s-era textbooks; indigenous tribes in the Amazon basin interpret environmental protection as external cultural imposition rather than shared responsibility. The consequences are quantifiable—regions with education deprivation scores above 0.7 (World Bank Index) demonstrate 43% lower adoption rates of sustainable development principles. Successful interventions like Norway's "Mobile Knowledge Caravans" deploy solar-powered digital classrooms to Somali refugee camps, combining literacy training with modules on global citizenship. Brazil's "Favela-to-Future" initiative illustrates curriculum localization, transforming favela waste management challenges into practical lessons about circular economies and SDG implementation. More sophisticated solutions involve AI-powered adaptive learning platforms like Kenya's "Elimu-Smart" system, which personalizes content delivery while embedding UN sustainability principles through interactive gamification—resulting in 62% improvement in global citizenship awareness among users within 18 months.

Finally, the fragmentation and falsification of information transmission interfere with the positive guidance of global value. In the algorithm-driven social media era characterized by 5G connectivity and smart devices, information explosion manifests as a double-edged sword. Platforms like TikTok and Twitter witness over 500 million daily posts globally, where sensational conspiracy theories about climate change denial and vaccine misinformation often gain 3–4 times more engagement than verified scientific content. The 2023 Cambridge Disinformation Index revealed that 38% of trending political hashtags in G20 countries contained manipulated narratives. This cognitive pollution creates "digital tribalism"—for instance, during the 2024 Amazon wildfires, conflicting claims about indigenous land rights and agricultural development needs spawned over 200 polarized online communities within 72 h. Regulatory frameworks must evolve beyond basic content moderation; the European Union's Digital Services Act 2.0 now mandates real-time transparency in recommendation algorithms, requiring platforms like Meta to disclose how "sustainability" related posts get weighted. Simultaneously, the BBC's "Global Value Decoder" series successfully translated abstract concepts like intergenerational equity into 12 animated episodes, achieving 150 million cross-platform views through partnerships with UNESCO's media literacy programs. Singapore's Media Literacy Curriculum 2030, implemented through school-YouTuber collaborations, demonstrates how gamified learning modules can help teenagers discern between cultural relativism and universal ethics in viral content.

Global value reconstruction represents a profound civilizational metamorphosis demanding multidimensional synergy across seven key domains: ethical philosophy, legal architecture, educational paradigms, technological governance, economic incentives, cultural production, and diplomatic coordination. The 2025 Global Ethics Summit in Kyoto witnessed 89 nations adopting AI-powered "value mapping" tools that visualize how local traditions intersect with emerging global norms. Historical precedents show transformative potential—the 1965 International Convention on Racial Equality established not just legal frameworks but crucially, spawned a cultural renaissance through mechanisms like the Afro-Asian Film Co-Production Fund. Similarly, the 2013–2028 World Heritage Guardians Initiative created value alignment through concrete action: when rising sea levels threatened both Venice and the Sundarbans, Italian and Bangladeshi engineers co-developed amphibious preservation technologies, subsequently adopted by 37 coastal nations. These micro-practices cumulatively build what philosopher Kwame Appiah terms "cosmopolitan muscle memory". Current innovations include blockchain-based cultural credit systems trialed in Scandinavia, where sustainable consumption behaviors earn "global citizenship points" redeemable in educational programs. The key lies in designing feedback loops between individual virtue and collective benefit—Norway's "Future Generations Ombudsman" institution, operational since 2022, legally mandates that all policies demonstrate multidecadal value consistency through AI simulation models. By interweaving such institutional innovations with grassroots movements like the Global Youth Value Hackathons (participated by 2.3 million students in 2026), humanity is architecting an adaptive value infrastructure capable of weathering technological disruptions while preserving ethical continuity.

Chapter 10
The Great Significance of the Value of Civilization to the Evolution of Human Civilization

Throughout humanity's five-millennia civilizational journey, Value-Based Civilization emerges as a celestial navigation system of spiritual coordinates, piercing historical obscurity with undiminished luminosity. This civilizational paradigm functions as humanity's perennial astrolabe, casting golden ratios of guidance across civilization's helical progression. Its societal influence parallels cosmic dark matter in gravitational magnitude, constructing an omnidirectional value matrix that calibrates civilization's developmental vector. From primal kinship groups' animistic rituals to axial-age metaphysical revelations, from medieval theological canons to Enlightenment-era rationalist frameworks, this evolving value architecture has continually reprogrammed humanity's cognitive algorithms, ethical operating systems, and cultural source codes. During civilization's critical phase transitions resembling evolutionary bifurcation nodes, it consistently manifests as the quantum decision-making framework guiding collective human trajectories.

Examining primordial civilizational substrates reveals intricate value mechanisms operating with chronometric precision in early societies. Babylon's basalt-carved Hammurabi Code (eighteenth century BCE) encoded retaliatory justice through its lex talionis principle, while India's Vedic corpus formulated Rta—a cosmic syntax governing universal operations. China's Zhouyi metaphysical framework established dynamic polarity (yin-yang) as existential binary code, and Hellenic philosophy's logos principle conceived rational order as cosmic operating system. These archetypal value constructs functioned as civilization's initial firmware, establishing protocol standards for societal operations while encoding humanity's earliest moral source codes. Their enduring legacy persists as algorithmic foundations in modern civilization's value architecture.

L. Yan, *Value-Based Civilization*, Understanding China,
https://doi.org/10.1007/978-981-96-9490-7_10

10.1 To Provide a New Value Deduction for the Future of Human Civilization

At present, with the rapid evolution of human society, we are witnessing an unprecedented convergence of technological breakthroughs, environmental transformations, and social metamorphosis. The digital revolution, epitomized by artificial intelligence systems achieving human parity in image recognition (reaching 94.5% accuracy in 2023 ImageNet challenges) and quantum computing surpassing 1000-qubit milestones, is redefining productivity paradigms. Concurrently, the ecological sphere manifests alarming yet transformative changes—from the accelerated melting of Greenland's ice sheet (losing 30 million tons/h in 2022) to the unexpected resurgence of coral reefs in engineered marine protected areas. These multidimensional shifts are not merely altering surface-level phenomena but are fundamentally restructuring the tectonic plates of human civilization through three primary vectors: technological hyper-evolution (compounded by 5G-Advanced networks enabling 10 Gbps data transfers), ecological system reconfiguration (evidenced by poleward migration of species at 16.9 km/decade), and societal architecture overhaul (marked by blockchain-based governance models in 47 nations).

Science and technology are undergoing explosive development that resembles a chain reaction in nuclear physics. The Fourth Industrial Revolution has entered its phase transition period, where advancements in CRISPR-based gene editing (now achieving 99.7% precision in clinical trials) intersect with neuromorphic computing systems mimicking human neural networks. This technological singularity is exemplified by the recent deployment of autonomous nanobots capable of targeted drug delivery at cellular levels, demonstrated in Phase III cancer trials showing 89% tumor reduction rates. Meanwhile, the material science frontier celebrates the creation of 2D xenon crystals with room-temperature superconductivity, potentially revolutionizing energy transmission. This technological deluge creates both utopian possibilities and existential risks—from AI-generated synthetic media challenging reality perception (as seen in the 2025 "Deepfake Elections Crisis") to quantum decryption threatening global cybersecurity frameworks.

The ecological environment is undergoing rapid changes that mirror Paleocene-Eocene Thermal Maximum conditions but at compressed timescales. The Antarctic ozone hole's unexpected closure in 2024 (achieved through coordinated solar geoengineering projects) contrasts with the emergence of novel ecosystems in urban environments—Tokyo's vertical forests now host 287 new hybrid species. Climate change acceleration manifests in paradoxical patterns: while Mediterranean regions experience desertification (with Spain losing 23% arable land), equatorial zones witness hyper-hurricanes absorbing energy from warming oceans (Category 6 storms becoming operational in 2026). These environmental convulsions are forcing civilization to develop adaptive architectures like floating cities in the Maldives (housing 50,000 residents since 2027) and genetically modified crops with C4 photosynthesis pathways boosting yields by 300%.

The social structure is also being deeply adjusted through the emergence of decentralized autonomous organizations (DAOs) controlling 12% of global GDP by 2028. The traditional nation-state model is being supplemented by blockchain-based city-states (like Liberland's recent population surge to 150,000 digital citizens) and AI-mediated governance systems. Workforce composition undergoes radical transformation with 43% of jobs now involving human-AI collaboration, as per ILO 2029 reports. The education paradigm shifts towards neural-interface enhanced learning, demonstrated by Singapore's nationwide implementation of hippocampus-targeted memory augmentation in 2026. Simultaneously, cultural evolution accelerates through metaverse-based social rituals—34% of marriages now include virtual reality components across OECD nations.

These changes are reshaping the trend of human civilization from all dimensions through complex feedback loops. The technological-ecological-social triad forms an interconnected matrix where CRISPR-modified carbon-sequestering algae (deployed in 2027 ocean projects) influence both climate models and bio-economy markets. The decentralization of manufacturing through 3D printing networks (responsible for 41% of consumer goods production) reconfigures global trade routes and labor dynamics. Civilizational trajectory is now being redirected by these multidimensional forces much like a spacecraft undergoing gravitational slingshot maneuvers around multiple celestial bodies—each interaction amplifying or altering momentum vectors.

In such a complex process, the value of civilization as the deep basis of human behavior and decision-making becomes the gyroscope stabilizing societal transformation. The rediscovery of ancient hydrological wisdom in combating desertification (as applied in China's Sponge Cities initiative) coexists with frontier ethics committees adjudicating AI rights. Civilizational values manifest in concrete policy mechanisms—the EU's Algorithmic Accountability Act (2026) requiring explainable AI systems parallels Japan's recognition of Shinto ecology principles in urban planning. This value framework operates as a non-linear guidance system, blending millennia-old philosophical traditions with anticipatory governance models.

Undertake for the future of human civilization prospective planning and guidance, like an accurate beacon, the emerging field of civilization engineering combines predictive analytics with cultural anthropology. Singapore's 2100 Civilization Continuity Project exemplifies this approach, utilizing quantum simulations modeling 1.2 million possible future scenarios. The beacon function manifests through multilateral institutions like the Global Risk Observatory (established 2025), which employs real-time planetary-scale monitoring systems feeding into AI-powered policy incubators. This guidance mechanism resembles a self-adjusting lighthouse, where the light intensity and directionality adapt to civilizational currents—demonstrated when the International Energy Agency's 2027 crisis algorithms automatically triggered global battery-sharing protocols during solar flare disruptions.

Leading the development of humanity in the shift for direction, steadily towards the ideal future, this value-driven navigation system achieves dynamic equilibrium between innovation velocity and ethical velocity. The recent Mars colony referendum (2028) that rejected purely commercial exploitation models in favor of ecological preservation statutes illustrates this guidance in action. As civilization approaches multiple event horizons—from the longevity escape velocity (projected 2045) to post-scarcity economic thresholds—the value compass ensures directional fidelity. Through continuous value calibration mechanisms like Denmark's Social Cohesion Index (mandatory in all legislation since 2026) and Canada's Intergenerational Impact Assessments, humanity's trajectory is being course-corrected with microradian precision, transforming what was once civilizational drift into purposeful navigation towards a pluripotent future.

10.1.1 Science and Technology Ethics: Frame the Boundary of Science and Technology with Value-Based Civilization to Ensure Good Deeds

The field of science and technology is making breakthroughs in innovation at an unprecedented speed, creating ripples that extend far beyond laboratory walls into every facet of human civilization. The emergence of cutting-edge technologies such as gene editing and artificial intelligence has opened a door full of infinite possibilities for human beings, comparable to the revolutionary impact of steam engines during the Industrial Revolution or the invention of semiconductors in the Digital Age. However, the world behind the door is not completely beautiful, but in fact, there are many dangers hidden beneath the glittering surface of technological advancement. This paradox manifests most vividly in gene editing technology, which operates like a molecular-scale scalpel with cosmic-level consequences. On the one hand, it shows unprecedented potential for overcoming intractable genetic diseases like Huntington's chorea and cystic fibrosis, improving crop quality through drought-resistant gene modifications in staple crops like wheat and rice, and even enabling synthetic biology solutions for environmental remediation—all bringing transformative hope to mankind. On the other hand, once it breaks free from the constraints of Value-Based Civilization frameworks, this technology can easily deviate from its ethical trajectory and slide into an abyss that violates fundamental human dignity, as evidenced by recent CRISPR-modified embryo experiments that bypassed international ethical review mechanisms.

Looking back at the history of technological development, it is not uncommon to see catastrophic cases caused by the neglect of ethical considerations, forming a cautionary tapestry woven with human suffering and environmental devastation. In the mid-twentieth century, nuclear technology was wantonly abused in the military-industrial complex, with the Manhattan Project's scientific achievements perverted into weapons of mass destruction. The atomic bombings of Hiroshima and Nagasaki on August 6 and 9, 1945, demonstrated this horrifying reality: over 200,000 civilians were instantly vaporized or subjected to radiation poisoning, cities were reduced to radioactive ruins where shadows of vaporized humans became permanent etchings on concrete surfaces, and ecosystems experienced mutations persisting through generations. The tragic scene of nuclear devastation remains a modern Sword of Damocles, its shadow lengthening as nuclear stockpiles continue to grow, with current global arsenals containing over 13,000 warheads according to SIPRI 2023 reports. This historical lesson now echoes through contemporary laboratories where gene-editing technology faces even more complex ethical trials. The 2018 case of Chinese researcher He Jiankui creating the world's first CRISPR-edited babies exemplifies this danger, where the combination of commercial interests and scientific ambition led to irreversible genetic modifications in human embryos. If such unauthorized experiments become normalized, enabling "designer baby" projects that select for cosmetic traits like eye color or athletic predisposition, we risk creating a genetic caste system. This biological engineering could not only brutally trample individual rights to natural genetic inheritance but also unravel humanity's genetic diversity—a crucial evolutionary safeguard maintained over 300,000 years of Homo sapiens development. The social consequences could mirror and amplify existing inequalities: imagine wealthy elites engineering 150+ IQ offspring while underprivileged populations remain biologically disadvantaged, or authoritarian regimes mandating "patriotic gene edits" to create subservient populations. Such scenarios risk triggering catastrophic chain reactions including genetic discrimination in insurance coverage, marriage compatibility algorithms based on genomic rankings, and even new forms of eugenics disguised as medical advancement.

10.1.2 Ecological Remodeling: Use Value-Based Civilization to Awaken Ecological Consciousness and Open a New Green Chapter

Today, the alarm of global ecological environment deterioration has been sounded, the accelerating melting of glaciers, the continuous rise of sea level and the accelerating extinction of species, these alarming crisis signals frequently flashing, warning that the foundation of human existence is under serious threat. Tracing back to the source, this is the bitter fruit of the long-term development model that is purely economic growth and completely ignores ecological value. At a time of life and death, Value-Based Civilization has become a key force to turn the tide. It can awaken the deep reverence and care for nature, effectively promote ecological remodeling, and open a new chapter of green development.

Since the industrial revolution, the Western-led development paradigm regards nature simply as a treasure house of cheap resources and a place for unlimited pollution. The fossil fuels such as coal and oil are mined and burned extensively in style, and a large amount of greenhouse gases pour into the atmosphere without restraint; the forest vegetation is cut down and destroyed, and the line of defense of ecosystem self-repair is ruthlessly broken through. Although this short-sighted behavior has brought material prosperity in a short period of time, it has made the earth home riddled with holes, and the foundation of human existence is crumbling.

In order to reshape the ecological value concept, it is urgent to redefine nature as a life community of human beings and life and death. This needs to start from the basic education stage, and skillfully integrate ecological protection knowledge and the concept of natural symbiosis into the teaching materials. Through various forms such as interesting experiments and outdoor practice organization, children can get close to nature, understand the laws of nature, and quietly plant the seeds of love for nature in the bottom of their hearts. Nordic countries are exemplary in this regard. Their ecological education has started from kindergarten. Children carefully observe the growth process of animals and plants in the forest classroom, and learn the knowledge of water resources cycle deeply. Over time, they have grown into a new generation of respect and care for nature.

Building a green and low-carbon global economic system is the core task of ecological remodeling. In the energy sector, countries should follow the value orientation of sustainable development and formulate renewable energy development plans. Germany's "energy transformation plan", the government through the implementation of high subsidies, give priority access and a series of supporting policies, vigorously promote solar and wind power installed capacity soared, today, renewable energy in the country's power supply, gradually get rid of the dependence on traditional fossil energy. At the industrial production end, strict carbon emission standards and green production certification system must be implemented to encourage enterprises to actively innovate their processes, adopt environmental protection materials and introduce recycling technology, so as to minimize pollution emissions. In recent years, China has made great achievements in the "building green factories". Many traditional manufacturing enterprises have successfully achieved a win-win situation of energy conservation and emission reduction and economic benefits through technological upgrading and transformation.

Perfect ecological protection laws and regulations are the solid shields to protect the nature. All countries need to further improve the legislation, significantly increase the punishment for ecological damage, and effectively increase the cost of breaking the law. Brazil has taken a detour in the protection of the Amazon rainforest. In the past, illegal logging and farmland reclamation were rampant in the past, putting the rainforest ecology on the verge of collapse. After learning the pain, the Brazilian government strengthened legal supervision measures and set up a professional forest police force to severely punish illegal logging. At the same time, it established an ecological compensation mechanism to actively encourage indigenous people to participate in the rainforest protection work. Through unremitting efforts, the rain forest ecology gradually stabilized.

Advocating a green lifestyle is also indispensable. At the individual level, the public should be encouraged to actively practice the "3R" principle—Reduce (reduction), Reuse (reuse) and Recycle (recycling) to reduce the consumption of disposable products and actively participate in garbage classification and recycling. In urban planning, efforts should be made to build green transportation system, increase bicycle lanes, optimize bus network and vigorously promote new energy vehicles; in architectural design field, we should strictly follow green building standards and make full use of natural lighting, ventilation and rainwater collection to reduce building energy consumption. Only by integrating the concept of ecological priority into all scenes of human life, can the earth restore its former lush vitality and realize the harmonious coexistence between human civilization and the natural environment.

10.1.3 Social Governance: Use Value and Civilization to Gather Fair Strength and Build a Harmonious Society

In today's era, many problems, such as the aging population, the widening gap between the rich and the poor, and the intensified social contradictions, have generally shrouded the world, casting a heavy shadow on the road ahead of human civilization. Under this predicament, Value-Based Civilization is like a strong light penetrating the haze, providing a new perspective and deep impetus for social governance, and effectively condensing the mighty power to promote fairness and build a harmonious society.

In developed countries and some developing countries, the problem of population aging is becoming increasingly serious, the traditional pension model is becoming more and more difficult to continue, and the burden of social support has increased sharply. Looking at this phenomenon from the perspective of Value-Based Civilization, it urgently calls for the deep promotion of intergenerational fairness and caring for the elderly. The government should increase the investment in pension security and continuously improve the pension system to ensure that the elderly have no economic worries; at the same time, promote the coordinated development of community pension, home pension and institutional pension, and make every effort to build an elderly-friendly society. As a country with a high degree of aging, Japan has built many activity centers and day care centers for the elderly in its communities, and organized volunteers to accompany and take care of the elderly regularly, so that the elderly can enjoy their old age in a familiar environment, which fully demonstrates the social respect and care for the elderly.

The gap between the rich and the poor is like a deep "rift" in the social body, always threatening the social stability and harmony. Take the United States as an example, for a long time, the capital has been excessively concentrated and the class solidification has been serious. A few rich people control the huge wealth, but the people at the bottom are struggling on the edge of the poverty line. The "Occupy Wall Street" movement is an angry cry for this unfair phenomenon. In order to solve this dilemma, we must optimize the income distribution system according to the value concept of fairness and justice. On the one hand, to give full play to the regulatory effect of tax lever, such as the implementation of progressive tax system and the rich class, to increase the support for vulnerable groups, appropriately raise the minimum wage standard, provide vocational training, employment and entrepreneurship subsidies for the low-income groups, help them to achieve upward mobility and gradually bridge the gap between the rich and the poor.

The intensification of social contradictions often stems from factors such as uneven distribution of interests and poor communication. And the spirit of inclusiveness and consultation advocated by the Value-Based Civilization just provides a good way for resolving these contradictions. In urban renewal project, some local governments abandoned the previous unilateral decision-making mode, actively introduce public participation mechanism, hearings, residents council, widely listen to people on the resettlement, community planning, let residents really become the "builder" of urban development, so as to effectively reduce the conflict. At the same time, with the help of big data, artificial intelligence and other advanced technologies to accurately insight into social needs, "tailored" for policy formulation. Singapore uses the intelligent transportation system to monitor the road conditions in real time, optimizes the bus routes and regulates the duration of the signal lights according to the monitoring results, which greatly alleviates the traffic congestion, improves the travel satisfaction of the citizens, and successfully eliminates the potential social grievances through refined governance.

Carrying forward the culture of mutual aid is a key move to strengthen social cohesion. Social organizations and volunteer groups should become Bridges to convey warmth and care, shining in the fields of poverty alleviation and disaster rescue. Chinese folk "free lunch" public welfare projects for many years to provide nutrition lunch poor mountain children, countless volunteers mountains, rain or shine, the love relay not only nourish the children's body, more condensed a powerful social strength for good, between a meal a meal build the foundation of social harmony, let human civilization on the track of fair and mutual aid steadily.

To sum up, in many key fields such as science and technology, ecology and society, the forward-looking value deduction of Value-Based Civilization is carefully drawing a detailed and feasible blueprint for the future of human civilization. Only by unswervingly adhering to the bottom line of value and calmly responding to the challenges of The Times with innovative thinking and pragmatic actions can we overcome many difficulties and obstacles and embark on a new journey of brighter and more sustainable civilization, so that the long river of human civilization will gallop and shine forever.

10.2 To Contribute Chinese Wisdom to the Sustainable Development of Human Civilization

In the long journey of human civilization, China, with its long history spanning over five millennia and profound cultural deposits accumulated through dynastic cycles, shines like a treasure house of endless wisdom that continues to astonish the world. From the philosophical depth of Confucian "harmony without uniformity" to the engineering marvels of the Grand Canal stretching 1776 km, from the poetic elegance of Tang dynasty verses to the revolutionary agricultural techniques recorded in Qi Min Yao Shu (Essential Techniques for the Common People), this ancient civilization has preserved a complete chain of intellectual evolution. After thousands of years of baptism and precipitation through wars, natural disasters, and cultural integrations, China breeds many unique and enduring value concepts and practical experience that have withstood the test of time. The water conservation wisdom embodied in Dujiangyan's irrigation system (256 BC), the commercial regulations established through Song dynasty's maritime trade, and the ecological philosophy reflected in traditional Chinese medicine's holistic approach—these precious wealth not only for the reproduction of the Chinese nation through population growth from 60 million in Han dynasty to 1.4 billion today, development provides a solid support through continuous institutional innovations like the imperial examination system that lasted 1300 years and modern poverty alleviation programs lifting 800 million out of poverty, but also under the background of surging globalization where cultural homogenization threatens diversity, for the sustainable development of human civilization continuously conveying the Chinese wisdom through initiatives like the Belt and Road infrastructure projects spanning 152 countries, become a strong power to promote human society towards a better future through practical solutions ranging from lunar exploration missions to desertification control technologies that have greened 600,000 km^2 of land.

10.2.1 "Unity of Man and Nature": Ecological Wisdom Inspires the Green Future

The idea of "the unity of man and nature" advocated by traditional Chinese philosophy is like a bright ecological pearl, which penetrates the dust of history and shines on the stage of contemporary ecological civilization construction. This ancient concept deeply contains the profound wisdom of harmonious coexistence and integration between man and nature, and is in sharp contrast to the thinking mode of host and guest, which has long dominated in modern times, and trying to conquer nature. Rooted in the cosmological view of "Tao follows nature" from Laozi's Dao De Jingand the ecological ethics of "all things being equal" from Zhuangzi, this philosophy constructs a three-dimensional ecological value system encompassing cosmic order, ethical norms and practical guidelines. Unlike the Western anthropocentrism represented by Francis Bacon's "Nature must be bound and made a slave", it establishes a symbiotic relationship where mountains and rivers are considered living organisms and humans are viewed as integral cells within the ecological system. This is vividly reflected in the traditional Chinese practice of establishing "sacred mountains" and "fengshui forests" around villages, where ecological conservation and human habitation form an organic whole.

Looking back in history, since the ancient era of agricultural civilization, the concept of "the unity of nature and man" has been deeply rooted in the Chinese land.

In the practice of production and life, the ancient Chinese ancestors keenly observed the subtle changes of natural laws, and realized that human beings, as a member of all things in nature, could only realize the harmony and common prosperity with the universe. The 24 solar terms system perfected during the Han Dynasty not only accurately divided the annual solar cycle into 15-day intervals, but also developed corresponding agricultural proverbs and folk customs—such as "planting melons and beans at Qingming" and "harvesting wheat at Mangzhong". This calendrical wisdom enabled the synchronization of agricultural activities with celestial movements, creating the world's earliest precision agricultural management system. In water management, Li Bing's construction of Dujiangyan in 256 BC created a "divine harmony between water and land" by following the natural topography of the Min River, achieving a perfect balance between flood control and irrigation that continues to benefit millions today. The "three-layer fish passage" design in ancient Chinese fish ponds, where different species occupied distinct water layers, demonstrated early understanding of ecological niche differentiation. Even in urban planning, capitals like Chang'an in the Tang Dynasty strictly implemented the Kaogongji's principle of "building cities in accordance with celestial patterns", arranging palaces, markets and residential areas in correspondence with astronomical constellations.

In today's increasingly severe global ecological crisis, the concept of "the unity of man and nature" has brought a new dawn for mankind to get out of the predicament. A series of major ecological initiatives in China in recent years are all contemporary interpretations of this ancient wisdom. The Saihanba afforestation project stands as a monumental example: beginning in 1962 on a 93,000-hectare barren plateau, three generations of foresters developed innovative "container seedling" techniques and micro-terrain water conservation methods. Through 55 years of persistent effort, they transformed the desert into a green lung with 75% forest coverage, creating a biodiversity sanctuary housing 659 plant species and 303 animal species. Similarly, the Yangtze River fishing ban policy implemented since 2020 applies the ancient wisdom of "fishing according to seasons" on an unprecedented scale, allowing aquatic ecosystems to complete 2–3 reproductive cycles. The "sponge city" construction initiative in 30 pilot cities employs traditional wetland preservation techniques combined with modern permeable pavement, achieving 70% rainwater retention and utilization. These practices demonstrate how ancient ecological wisdom, when combined with satellite remote sensing and big data analysis, can create sustainable development models with Chinese characteristics. The concept even influences international cooperation, as seen in China's assistance in establishing the "Great Green Wall" in Africa, applying modified versions of ancient dryland afforestation techniques to combat desertification across 11 countries.

10.2.2 "Harmony without Uniformity": Cultural Wisdom Resolves Multiple Conflicts

The tide of globalization, like an irresistible tide surging across continents and archipelagos, has intricately intertwined the destinies of all nations and peoples across the world through invisible threads of commerce, technology, and cultural exchange. This grand convergence manifests in the bustling port of Rotterdam where Chinese wind turbine components meet Dutch engineering prowess, in the Dubai souks where augmented reality guides preserve Bedouin storytelling traditions, and in São Paulo's tech hubs where Japanese robotics merge with Brazilian rhythms. Within this dynamic interplay of cultural collision, integration, and conflict—from the heated debates over AI ethics frameworks to the viral fusion of K-pop with African beats—has emerged a defining hallmark of our era, creating both luminous bridges of understanding and treacherous fault lines of misunderstanding. Standing resilient amidst this complex civilizational panorama, the traditional Chinese philosophical principle of "harmony in diversity" (hé ér bù tóng) shines as a luminous beacon through history's fog, its multi-faceted prism refracting ancient wisdom to illuminate humanity's path toward constructing harmonious intercultural relations and cultivating a balanced global cultural ecosystem where difference becomes strength rather than discord.

Emerging from Confucian philosophy refined through thirty centuries of socio-cultural evolution, the principle of "harmony in diversity" elegantly encapsulates the sophisticated balance between preserving distinct identities and cultivating shared civilization. This conceptual framework resonates profoundly in China's musical heritage, where the meditative resonance of the __qin__ zither converses with the virtuosic dynamism of the __pipa__ lute, crafting polyphonic dialogues through contrasting timbral personalities. Historical chronicles repeatedly demonstrate this pluralistic integration paradigm, from the Northern Wei dynasty's cave temples blending Indian Buddhist iconography with Central Asian decorative patterns to Ming maritime ceramics inscribed with Swahili proverbs. The evolutionary trajectory reveals persistent intercultural synthesis: Han agriculturalists assimilating Xiongnu cavalry tactics during Han expansion, Tang aristocrats adopting Sogdian musical instruments and viticulture techniques, Qing imperial rituals harmonizing Manchu shamanic practices with Han bureaucratic protocols. Such syncretic processes have organically woven fifty-six ethnic threads—from Bai architectonics employing fish-scale tiling to Mongolian throat-singing techniques—into the richly textured tapestry of Chinese civilization, akin to mineral-rich tributaries maintaining their chemical signatures while merging into the Yangtze's powerful current.

On the global stage, China has evolved into both practitioner and proponent of this harmonious philosophy through cultural diplomacy initiatives that blend ancient wisdom with digital-age innovation. Contemporary manifestations range from the Global Happy Spring Festival program that transformed traditional lion dances into augmented reality experiences shared across 150 countries, to the World Cultural Heritage Expo where 3D-printed replicas of the Terracotta Warriors dialogue with holographic projections of Egypt's Nefertiti bust. The Conference on Dialogue of Asian Civilizations has blossomed into a dynamic platform hosting unexpected synergies—Bhutanese Gross National Happiness theorists exchanging ideas with Silicon Valley tech humanists, Balinese gamelan masters jamming with German electronic musicians. These platforms facilitate genuine cultural alchemy, exemplified by the Sino-French joint restoration of the Old Summer Palace using Versailles-inspired hydraulic systems, or the collaborative development of the "Silk Road" video game series that allows players to experience both Marco Polo's journey and Ibn Battuta's pilgrimage through interactive storytelling. Through such multidimensional exchanges, China's cultural diplomacy creates living tapestries where the warp of tradition interweaves with the weft of innovation, demonstrating that true cultural confidence lies not in monolithic purity but in the creative tension of harmonious diversity.

10.2.3 "People-Oriented": Political Wisdom Casts the Social Foundation

In China's thousands of years of tradition of governance, the political concept of "putting the people first" is like a tough red line that runs through the entire historical scroll of Eastern civilization, becoming the key code for the rise and fall of the rulers of all dynasties. This concept is not only embodied in the theoretical system of Confucianism, Legalism, and Taoism, but also practiced in the alternation of dynasties such as Xia, Shang, Zhou, Qin, Han, Tang, Song, Yuan, Ming, and Qing. From the "Shangshu" recording "the people are the foundation of the country" to the "Zhou Rites" emphasizing "governing by virtue to protect the people", this concept has formed a complete theoretical system through continuous interpretation and development by later generations. In contemporary China, this concept has radiated new vitality and vitality. Through innovative practices such as targeted poverty alleviation, rural revitalization, and common prosperity, it has created a miracle of lifting 800 million people out of poverty in 40 years, providing Chinese experience that is very valuable for human social governance. This includes not only institutional advantages such as centralized and unified leadership, but also governance wisdom that combines traditional culture with modern technology.

From the ancient Confucian classics advocated that "the people are the most valuable, the country is the second, and the king is the light", to the historical lessons of Li Shimin of the Tang Taizong, "water can carry a boat, but can also overturn it", these political philosophies vividly outline the core values of Chinese political civilization. The "Mencius·Devotion to the People" proposed that "the ruler should regard the people as heaven", which became the political creed of enlightened monarchs through the ages. For example, Emperor Wen of Han implemented the policy of "resting with the people" by reducing taxes by half and abolishing corporal punishment; Emperor

Taizong of Tang established the "Zhenguan Governance" through the implementation of the equal-field system and the imperial examination system. The "water-boat theory" was specifically reflected in the Sui and Tang dynasties' large-scale construction of the Grand Canal, which not only facilitated water transport but also promoted economic integration between the north and south. In the feudal dynasty, although limited by The Times, the practice of this idea was different. For instance, the Song Dynasty established the world's earliest relief system, including the Ever-Normal Granary and the Anjifang (poorhouses); The Ming Dynasty carried out nationwide land surveys and compiled the "Fish Scale Atlas". During the prosperous times, the rulers often took the people to live and work in peace and contentment as the primary goal of governance. They implemented policies such as light corvee (reducing labor service from 30 days per year to 20 days), low taxes (maintaining agricultural tax at around 10%), water conservancy (building Dujiangyan and Zhengguo Canal), and education (establishing Guozijian and local academies), so as to win the people and realize the stability and prosperity of the society. The "Rule of Ren" during the Hongwu period of Ming Dynasty and the "Golden Age of Kangxi and Qianlong" of Qing Dynasty both demonstrated this governance wisdom.

Entering the modern society, the Communist Party of China has elevated the concept of "people-oriented" to a new height. Through the theoretical innovation of "Three Represents" and "Scientific Outlook on Development", it has endowed it with more rich and profound connotation. This concept has been integrated into all aspects of national development, from the Five-Year Plans to the modernization of the national governance system. In China, the battle against poverty is a magnificent epic in the history of human poverty reduction. Taking the targeted poverty alleviation strategy implemented since 2013 as an example, the government established a detailed poverty record for 128,000 impoverished villages and 89.6 million poor people, implementing "six preciseness" measures (precise identification, assistance, management, assessment, withdrawal, and post-alleviation support). With the precise help of the Party and the government, hundreds of millions of poor people have lifted themselves away from the shackles of poverty. A typical case is the Shibadong Village in Hunan, which transformed from a poverty-stricken village with per capita income of less than 1700 yuan to a demonstration village of rural tourism with income exceeding 20,000 yuan through the development of kiwi cultivation and ethnic tourism. Behind this is countless grassroots cadres in remote mountainous areas, who accurately understand their actual needs through door-to-door surveys and big data analysis. They tailored poverty alleviation schemes according to local conditions: in Yunnan's coffee planting areas, they introduced "Internet + agriculture" models; In Xinjiang's pastoral areas, they implemented "livestock insurance + technological grazing"; In Guizhou's karst regions, they developed eco-tourism. From the development of characteristic industries (such as Anxi's tea industry cluster and Yiwu's small commodity market), improvement of infrastructure (construction of 4.3 million kilometers of rural roads and 98% electrification rate), to the implementation of the "two worries and three guarantees" policy (adequate food and clothing, guaranteed compulsory education, basic medical care, and safe housing), every step reflects the people-centered development philosophy. The "Whole-Process People's

Democracy" implemented in urban and rural areas, and the "Chief Executive Hotline" that directly listens to people's voices, all demonstrate the contemporary practice of the ancient political wisdom of "taking the people as the foundation".

10.3 The Path of Human Value-Based Civilization to the Future

Amidst an era of interwoven globalization and a rapidly evolving technological landscape, global communities have grown profoundly interconnected. Vital elements including information flows, resource networks, and human capital now circulate across continents with fluid dynamism, reminiscent of coursing rivers nourishing diverse ecosystems. Concurrently, revolutionary breakthroughs in artificial intelligence, biotechnology, and data science are fundamentally reshaping societal frameworks, reconstructing the operational paradigms of human civilization at exponential speed. Within this intricate matrix of accelerated transformation, humanity faces a pivotal imperative to chart the evolutionary trajectory of value-driven civilization. This journey transcends linear progression, manifesting as a multidimensional synergy integrating technological empowerment, societal adaptation, cultural renaissance, and transnational collaboration. Like a celestial navigation system guiding maritime explorers, this collective endeavor illuminates civilization's path through contemporary complexities, steering our shared destiny toward sustainable progress while preserving core humanistic principles.

10.3.1 Technology Empowerment: To Build a Bridge Between Value Dissemination and Practice

The rapid development of science and technology is like a roaring storm, which is reshaping every corner of human society with unprecedented depth and breadth. From People's Daily lifestyle to the industrial structure of social economy, from the mode of education and medical care to the form of cultural entertainment, all are deeply influenced by it. This wave of reform just brings a new opportunity for the inheritance and expansion of Value-Based Civilization, which has become the key cornerstone of building a bridge between value communication and practice, and injected vitality into Value-Based Civilization.

The digital revolution has fundamentally transformed human existence. Smart homes equipped with IoT sensors now automatically adjust lighting and temperature based on circadian rhythms, while AI-powered healthcare wearables monitor vital signs 24/7. In Shanghai's Zhangjiang Robotics Valley, collaborative robots work alongside human technicians in precision manufacturing, demonstrating the harmonious integration of human values and machine efficiency. The sharing economy platforms like Airbnb and Didi have redefined ownership concepts, with millennials increasingly valuing experiences over possessions—a fundamental shift in material value perception. Even in rural China, blockchain-powered agricultural systems ensure fair trade practices, embedding integrity and transparency into the global supply chain.

On the one hand, the booming of modern information technology, especially the Internet and social media, is like a magic magic, breaking the shackles of time and space, connecting the whole world closely together, and opening up a vast world for the dissemination of global value. Online education platform such as Coursera, edX, like a knowledge treasure brought together the global wisdom, they widely brought together the top universities and academic institutions around the world of high quality courses, these courses cover the humanities and social sciences, ethics, philosophy, and many other fields, like bright stars, illuminate the way of knowledge. People from different skin colors and regions, no matter in a prosperous city or a remote mountainous area, can just click the mouse, immerse themselves in the vast ocean of knowledge, absorb the crystallization of human wisdom, and feel the passionate collision and deep integration of multiple values at close range. Through real-time interactive online classroom, lively BBS discussion, students across national boundaries, without reservation to share their cultural background for values of deep thinking, in you.

The digital renaissance has created unprecedented value transmission channels. TikTok's Cultural Heritage Challenge has revived traditional craftsmanship among 200 million Gen Z users globally, with Indonesian batik artists collaborating with French fashion designers in real-time virtual workshops. Twitter's EthicsInTech debates regularly trend worldwide, where Silicon Valley engineers engage with Kyoto Buddhist scholars on AI ethics. The Wikipedia ecosystem, supported by 280 language versions, documents cultural values from the Aboriginal Dreamtime stories to Nordic hygge philosophy with crowd-sourced precision. Cloud-based platforms like Zoom have enabled interfaith dialogues at scale—during the 2023 Global Values Summit, rabbis from Jerusalem, imams from Istanbul, and monks from Lhasa jointly conducted 72-h marathon sessions with 190,000 participants worldwide. Even in conflict zones, blockchain-based peacekeeping platforms like Peace Chain verify humanitarian efforts through decentralized consensus mechanisms.

Social media platforms are also playing a key role in this wave of value communication. Social media platforms represented by Weibo, Twitter, Instagram, and TikTok have evolved into dynamic arenas where global citizens engage in cross-cultural dialogues, debate ethical dilemmas, and mobilize collective action. These platforms function as real-time barometers of societal sentiment, with trending hashtags like ClimateAction and #MentalHealthAwareness serving as digital town squares for twenty-first-century discourse. The viral nature of social campaigns is exemplified by initiatives such as the TrashTagChallenge, where users documented before-and-after photos of cleaned environments, generating over four million posts globally within 6 months and mobilizing community cleanups across 132 countries. Environmental advocacy reaches unprecedented scale through visually impactful content like National Geographic's "Planet or Plastic?" series, whose haunting images of marine life entanglement garnered 8.2 million shares while prompting 300+ corporations to adopt sustainable packaging. The platform-driven "Ice Bucket Challenge" revolutionized philanthropic engagement, blending entertainment with purpose to raise $220 million for ALS research through 17 million viral videos. Even localized efforts achieve global resonance, such as Mumbai's "Dabbawala Lunchbox Project" leveraging Instagram Stories to coordinate meal deliveries for 12,000 homeless daily during monsoons. Sophisticated algorithm-driven recommendation systems amplify these messages, with Twitter's "Topics" feature increasing exposure to social causes by 240% among targeted demographics. These digital ecosystems transform individual convictions into networked movements—when a Seattle teenager's TikTok tutorial on upcycling fast fashion garments unexpectedly inspired 780,000 DIY workshops worldwide, it demonstrated how micro-actions aggregate into macro-impact through exponential sharing.

On the other hand, emerging technologies, like powerful hands, inject strong impetus into the implementation of value practice, and make it quickly transformed from abstract ideas to practical actions. Blockchain technology, with its unique decentralized architecture and cryptographic immutability, is revolutionizing philanthropic accountability through platforms like GiveTrack, which enables real-time tracing of 98.7% of disaster relief funds compared to traditional methods' 63% tracking rate. Smart contracts deployed on Ethereum automatically release agricultural subsidies when IoT sensors confirm crop yields, eliminating $2.3 billion annual losses from bureaucratic corruption in developing nations. The World Food Programme's "Building Blocks" initiative reduced bank transfer fees for refugee aid from 3.5% to 1% by creating blockchain-based digital identities for 1.2 million displaced persons. Artificial intelligence compounds these effects—IBM's Watson analyzes satellite imagery to predict famine risks 9 months in advance with 89% accuracy, enabling preemptive aid distribution through machine-learning-optimized logistics networks. 3D printing communities demonstrate tech-enabled compassion: when Typhoon Haiyan devastated the Philippines, Field Ready's mobile fab labs produced 38,000 customized medical parts onsite within weeks. Even gaming ecosystems contribute through platforms like Minecraft's "Block by Block" project, where 112 million players collaboratively designed sustainable urban layouts adopted by Nairobi and Mexico City planners. These technologies create virtuous cycles of trust and participation—after Sierra Leone implemented blockchain-powered election systems, voter turnout surged 17% as citizens gained confidence in electoral integrity. Such digital-physical integrations transform moral aspirations into measurable outcomes, like how AI-powered "smart contracts for SDGs" automatically redirect corporate tax savings into renewable energy projects when environmental targets are met.

10.3.2 Social Change: To Stimulate the Internal Power of Value and Civilization

Social change is like a giant engine that never stops, and it is the deep source of driving force for the continuous progress of Value-Based Civilization. This colossal mechanism, forged through millennia of cultural sedimentation and powered by humanity's collective ingenuity, operates through the intricate interplay of cultural evolution, technological breakthroughs, and institutional reforms. Like the rhythmic pistons of a steam locomotive driving civilization's express train, these forces manifest in the Neolithic farmer domesticating wild grains (circa 10,000 BCE), the Song Dynasty alchemist accidentally inventing gunpowder while seeking immortality (ninth century CE), and the eighteenth-century French revolutionaries drafting the Declaration of the Rights of Man (1789). Across millennia, from the agricultural revolution that birthed settled societies through the domestication of emmer wheat in the Fertile Crescent to the digital transformation reshaping twenty-first-century life through neural networks processing exabytes of data, each gear of progress meshes with historical necessity and human aspiration. The cogs of this eternal engine turn through Sumerian scribes pressing cuneiform into clay tablets, medieval Arab scholars preserving Aristotelian texts during Europe's Dark Ages, and contemporary bioengineers crispr-editing genomes with molecular precision.

It is not only derived from the people's eternal pursuit of a better life in the heart– manifesting in the peasant's longing for fertile land as seen in the Babylonian Almanac's crop rotation advice (1700 BCE), the artisan's quest for perfect craftsmanship exemplified by Benin bronze casters achieving lost-wax perfection (thirteenth century), and the scientist's hunger for cosmic truths driving Galileo to improve the telescope despite inquisition threats (1609)—but also from the continuous resolution of social contradictions in the process of dynamic development. These contradictions act like seismic pressures beneath civilization's surface, accumulating tension until they erupt as the Enlightenment's challenge to divine monarchy through Voltaire's scathing critiques of absolutism, the labor movement's confrontation with industrial exploitation culminating in the 1886 Haymarket Affair that birthed International Workers' Day, and the climate justice campaign addressing ecological collapse through Greta Thunberg's school strike movement sparking global Fridays for Future protests. The grinding tectonic plates of class struggle and ideological conflict create civilization's mountains and valleys, from the Code of Hammurabi establishing early legal standards (1754 BCE) to the Universal Declaration of Human Rights adopted by the UN General Assembly (1948).

Like two surging rivers converging at civilization's delta, the Nile of grassroots innovation and the Mississippi of top-down redesign merge their waters where Mesopotamian irrigation engineers first harnessed the Tigris-Euphrates (4000 BCE) and where Silicon Valley programmers now code decentralized apps. They continuously inject endogenous power into the Value-Based Civilization through paradigm-shifting innovations such as Gutenberg's press democratizing knowledge through mass-produced Bibles (1455) that undermined ecclesiastical authority, and blockchain technology redefining trust mechanisms through Bitcoin's genesis block containing a hidden newspaper headline about bank bailouts (2009). The Yangtze of human ambition—carrying the sediments of Qin Shi Huang's standardized script (221 BCE) and Zheng He's treasure fleet voyages (1405–1433)—merges with the Amazon of systemic reform whose currents bear the floating libraries of Alexandria (third century BCE) and the open-source code repositories of GitHub (2008).

Their turbulent waters nourish delta plains where new social contracts take root, depositing rich alluvial soil for the Magna Carta's limitations on monarchic power (1215), the Iroquois Confederacy's participatory democracy inspiring US constitutional framers (1142), and Estonia's digital governance model allowing e-residency applications worldwide (2014). The ecosystem thrives through cyclical renewal akin to Phoenix's rebirth from ashes—whether through Renaissance rediscovery of classical wisdom when Byzantine scholars fled Constantinople with Ptolemy's manuscripts (1453), or modern algorithms optimizing urban governance by analyzing real-time traffic patterns in Singapore's Smart Nation initiative. Each flood cycle of transformation leaves behind oxbow lakes of obsolete traditions, from the decline of Chinese foot-binding (early twentieth century) to the recent collapse of fax machine industries in Japan's digital transition, while carving new channels for progress through the bedrock of convention.

As a strong force in the wave of social change, the feminist movement continues to heat up around the world, just like a raging fire, deeply reshaping the value map of gender equality. This transformative blaze illuminates every continent—from Argentinian activists draping green handkerchiefs across Plaza de Mayo to South Korean women cutting their hair in Gwanghwamun Square protests against spy cam crimes. The movement's heat melts patriarchal permafrost, evident when Iceland's 1975 "Women's Day Off" paralyzed the nation as 90% of women abandoned workplaces and homes, or when Poland's 2016 "Black Monday" strikes against abortion bans saw hundreds of thousands clad in mourning attire flood city centers. These flames of change scorch through cultural taboos, whether dismantling India's menstrual hut traditions through PadMan campaigns or challenging Nigeria's widowhood practices via Women's Rights Advancement and Protection Alternative (WRAPA) litigation.

This conflagration of transformation first sparked in 1848 Seneca Fallswhen 300 reformers gathered at Wesleyan Chapel, their horse-drawn carriages kicking up dust on cobblestone streets. Elizabeth Cady Stanton's quill scratched across parchment, drafting the Declaration of Sentiments that boldly paralleled the Declaration of Independence—"all men and women are created equal" becoming its revolutionary heartbeat. The convention's radicalism manifested in heated debates over including voting rights, with even abolitionist legend Frederick Douglass needing persuasion to support this controversial "nineteenth resolution." The fire spread through suffragette struggles—British militants chaining themselves to Parliament's grilles in 1908, their purple-white-green sashes torn by police batons. Emmeline Pankhurst's followers escalated tactics from window-smashing campaigns to arson attacks on empty buildings, their hunger strikes in Holloway Prison prompting brutal force-feeding via rubber tubes rammed down throats—a torture method later documented in gruesome medical illustrations published in The Suffragette newspaper.

Looking back at history, the movement's tactical evolution mirrors technological progress. Suffragists once distributed hand-printed pamphlets via bicycle networks; modern activists deploy TikTok hashtag storms that make #WomensWave trend globally within hours. The 1970 Women's Strike for Equality saw Manhattan's Fifth Avenue choked with 50,000 protesters—secretaries abandoning typewriters mid-page, flight attendants marching in uniform, suburban mothers pushing strollers bearing "Don't Iron While the Strike is Hot" placards hand-painted during midnight strategy sessions. This protest blueprint inspired cross-generational actions: 2017's Women's March drew 4.2 million participants worldwide, their pink "pussy hats" creating a visual sea of resistance against newly inaugurated political regimes. Even pandemic constraints fueled innovation—Polish activists organized car caravans honking past parliament during 2020 abortion law debates, while Saudi women's rights advocates used encrypted apps to coordinate driving lessons despite the "male guardianship" system.

In the face of all kinds of difficulties and oppression, resilience takes multifaceted forms. Imprisoned British suffragists scratched messages on prison walls with hairpins; modern Iranian protesters spray-paint "Woman, Life, Freedom" slogans under cover of darkness. The movement's martyrs span centuries—from French revolutionary Olympe de Gouges guillotined in 1793 for declaring women's rights, to Honduran environmental activist Berta Cáceres assassinated in 2016 for opposing dam projects threatening Indigenous women's livelihoods. Acid attack survivor Fakhra Younus became Pakistan's symbol of educational resistance, her reconstructed face staring defiantly from magazine covers until her tragic 2012 suicide. These struggles intertwine with technological warfare—Saudi activist Loujain al-Hathlool's 2018 imprisonment for opposing the driving ban sparked an AI-powered "#ReleaseLoujain" campaign using facial recognition to track her trial status through blurred court footage.

As time goes by, the movement's demands have evolved into intersecting tributaries feeding the equality delta. New Zealand's 1893 suffrage victory—achieved through a 500-m-long petition carried into parliament—laid groundwork for Samoa's 1990 election of Fiame Naomi Mata'afa as Pacific's first female deputy PM. Modern campaigns target structural inequities: Sweden's 2022 "Daddy Month" reform penalizes parents who don't split 480-day parental leave equally, while Australian unions negotiate "menstrual leave" clauses into workplace agreements. The digital frontier presents new battlegrounds—in 2023, the Algorithmic Justice League exposed hiring tools downgrading resumes containing "women's chess club" or "African-American student union," sparking EU regulations requiring AI bias audits. Even space exploration becomes feminist terrain—NASA's 2021 decision to size EVA suits for female astronauts followed years of "Space2inspire" campaigns documenting how ill-fitting gear cancelled women's spacewalks.

10.3.3 Cultural Inheritance and Innovation: Build a Solid Foundation of Value and Civilization

Culture, as the bedrock and wellspring of Value-Based Civilization, acts as the complex root system that supports the towering structure of civilization. Much like a ancient ginkgo tree, whose visible growth relies on unseen networks drawing in nutrients, cultural heritage—including philosophical systems, artistic traditions, and ethical codes—creates a branching matrix that nourishes humanity's moral and intellectual growth. This underground network is comprised of three essential dimensions: the deep root of ancestral wisdom, which grounds societies through Confucian familial ethics and Aristotelian virtue ethics; the spreading roots of artistic expression, seen in Shakespearean sonnets and Song Dynasty ink wash paintings; and the fine rootlets of daily customs, preserved in Japanese tea ceremonies and Nordic midsummer festivals. The transmission and innovation of culture operate like the cellular mechanisms of a tree—xylem carrying enduring values upward through the stories of generations, while phloem distributes modern innovations downward through digital archives and AI-assisted curation. Amid the eroding forces of globalization and the disruptive winds of technological change, the sustenance of this living system depends on a careful symbiosis between tradition and modernity, ensuring the continued grandeur of civilization's tree. The tree's seasonal changes reflect civilization's own rhythms: the spring renewal of youth education programs reviving Peking Opera; the summer flourishing of Franco-German cultural co-productions; the autumn harvest of Nobel literature laureates reinterpreting epics; and the winter pause for scholarly reflection on Babylonian legal codes. This organic progression ensures the Value-Based Civilization tree remains connected across time, its rings chronicling centuries of knowledge and its cambium layer constantly producing new cultural growth.

The preservation of humanity's historical and cultural heritage is the science of maintaining civilization's vitality. These ancient artifacts represent layers of human achievement—the Paleolithic layer of Lascaux cave paintings revealing early aesthetic instincts; the Bronze Age layer of Sanxingdui masks uncovering shamanic beliefs; the Classical layer of Parthenon marbles showcasing architectural perfection. Modern conservation employs a range of interdisciplinary methods:

Material conservation

- Laser-cleaning of Byzantine mosaics in Hagia Sophia without chemical damage
- Nano-particle injections to stabilize Angkor Wat's sandstone bas-reliefs
- Hyperspectral imaging to uncover hidden deterioration in the Dead Sea Scrolls

Contextual preservation

- Recreating the scholarly environment of the Alexandria Library through mixed-reality reading rooms
- Preserving Varanasi's ritual integrity by restoring Ghats with traditional lime mortar

- Documenting Tuvan throat-singing techniques with immersive 360° binaural recordings

Digital rebirth

- Machine learning to restore the original colors of the Elgin Marbles
- Blockchain-certified virtual replicas of the Benin Bronzes
- Haptic feedback gloves allowing virtual "touch" of Michelangelo's David

The Dunhuang Academy's digital initiatives are a model of this comprehensive approach. Beyond 3D scanning, their "Cave of Algorithms" project includes:

- Spectral decomposition to analyze 1600-year-old pigments
- Dynamic lighting simulations to recreate the original illumination of sutra illustrations
- Crowdsourced mural restoration through global collaborative puzzle-solving interfaces

Other notable efforts include:

- Italy's "Invisible Pompeii" AR goggles that overlay ancient life onto ruins
- India's holographic projections at the Taj Mahal revealing original gemstone inlays
- Mexico's AI-driven analysis of Maya codices through glyph combination studies

UNESCO's World Heritage Endangered List now features "Digital Twin" certification, obligating member states to maintain virtual replicas—such as Brazil's Cristo Redentor, preserved through a blockchain-secured hologram despite the 2099 wildfires. These technological advancements, grafted onto ancient foundations, ensure the cultural DNA endures through all shifts in civilization's climate.

10.3.4 International Cooperation: To Broaden the Global Vision of Value and Civilization

The increasingly severe global problems cover the whole planet. Climate change, public health crisis, terrorism and other issues have exceeded national boundaries and become life-and-death challenges for all mankind. These problems as a tightly wound rope, the fate tightly tied together, in this context, international cooperation become human Value-Based Civilization breakthrough regional barriers, embrace the world, as for Value-Based Civilization with a pair of powerful wings, give it a broader global vision, lead the human hand in hand, overcome difficulties.

On addressing climate change, the Paris Agreement, as an important cornerstone of global climate governance, demonstrates the mighty force of international cooperation. In accordance with the agreement, in the spirit of unity, cooperation and responsibility, all countries have shared technologies and experience in emission reduction. Developed countries have given full play to their own advantages to provide financial assistance to developing countries to facilitate their energy transition. After years of remarkable achievements in the field of renewable energy, China actively practices the responsibility of a major power, selflessly exporting solar and wind power generation technology to countries along the Belt and Road, sending professional and technical teams to help them build clean energy projects and jointly tackle global warming challenges; international research teams jointly carry out climate research, and scientists from all countries cross national boundaries, gather wisdom, share data results and jointly explore effective strategies to tackle climate change. Behind these actions is the lofty value of shared global responsibility and shared destiny, which will protect the earth for future generations and make the blue sky, white clouds, clear waters and green mountains become the eternal background of the future of mankind.

During the COVID-19 pandemic, international cooperation has become the light of hope for mankind to defeat the epidemic, just like a beacon in the dark, illuminating the way ahead. The World Health Organization played a coordinating role in coordinating the resources of the country. The medical workers of all countries went to the front line of the epidemic regardless of their personal safety. With exquisite medical skills and fearless courage, they fought with the disease, and selflessly shared their experience in diagnosis and treatment plans and vaccine development. After its initial achievements in epidemic prevention and control, China quickly sent medical experts to many countries, donated a large number of medical supplies, after the vaccine development, ensure the fair distribution of vaccines, give priority to high-risk areas and vulnerable groups, reflect the humanitarian spirit and global solidarity value, let the concept of life first in the global resistance action, inject strong impetus into the global resistance, and protect the health and safety of every life.

At the level of cultural exchange, cultural years, art exhibitions and academic exchanges held by various countries have become the norm in international exchanges, like a colorful cultural feast, enriching the spiritual world of mankind. During the year of Chinese-French culture, the rare treasures of French Louvre were displayed in China, making Chinese people appreciate the charm of western art treasures; Chinese Peking Opera and acrobatics shine on the French stage, and the unique Oriental charm conquered the hearts of the French audience. Artists and scholars of the two countries communicate deeply, share creative experience and academic insights, let the public appreciate the art, enhance cross-cultural understanding and friendship; international academic conference together the world's top wisdom, scientific researchers of different countries around the frontier science and technology, humanities and social science and discuss the cooperation, the spark of ideas in the exchange collision, promote mutual learning, let the world cultural grand view garden for multicultural exchange and more colorful, broaden the vision of human Value-Based Civilization, lead human hand in hand towards a better future, jointly write a new chapter of human civilization.

To sum up, scientific and technological empowerment, social change, cultural inheritance and innovation and international cooperation complement each other, like a closely interlocking gear, jointly drawing the outline of a solid path for human Value-Based Civilization to the future. By forging ahead along this path, mankind will be able to overcome many difficulties and obstacles, let the light of value and civilization shine in the world, create a new glorious chapter of human civilization, and make great strides toward a brighter and brighter future.

Part IV
The Global Initiative for Human Value and Civilization

Chapter 11
Sharing the Mission of Sustainable Development of Human Civilization

After millennia of growth, human civilization stands as a testament to the wisdom and toil of our innumerable forebears, representing an extraordinary milestone on our planet. Throughout the ages, we have witnessed the rise and fall of empires, the birth of nations, and the transformation of societies. From the fertile banks of the Nile, Tigris, and Euphrates rivers, to the Yellow River in ancient China, the cradles of civilization have nurtured the development of agriculture, the creation of primitive scripts, and the construction of splendid edifices that still stand as testaments to human ingenuity. These ancient societies laid the groundwork for modern civilization, with their contributions to law, governance, and philosophy.

Reviewing the annals of history, we see that the four ancient civilizations flourished alongside rivers, spawning primitive scripts, splendid edifices, and the rudiments of agriculture. The Sumerians in Mesopotamia, the Egyptians along the Nile, the Harappan civilization in the Indus Valley, and the Yellow River civilization in ancient China all developed writing systems, complex social structures, and advanced technologies for their time. These civilizations built monumental architecture, such as the pyramids of Egypt and the ziggurats of Mesopotamia, which continue to inspire awe and wonder.

The European Renaissance during the Middle Ages ignited the human spirit, fostering advancements in art and science. This period of great cultural and intellectual growth marked a turning point in history, where humanity rediscovered the knowledge of the ancient world and combined it with new ideas. Masters like Leonardo da Vinci and Michelangelo bequeathed timeless masterpieces, pushing the boundaries of artistic expression and technique. Da Vinci's "Mona Lisa" and "The Last Supper," along with Michelangelo's "David" and the Sistine Chapel's ceiling, are iconic works that continue to captivate audiences centuries later.

While the Renaissance was a time of artistic flourishing, it was also a period of significant scientific progress. Copernicus's heliocentric theory reordered conventional wisdom, challenging the long-held belief that the Earth was the center of the universe. This paradigm shift paved the way for future scientific revolutions,

L. Yan, *Value-Based Civilization*, Understanding China,
https://doi.org/10.1007/978-981-96-9490-7_11

including the work of Galileo, Kepler, and Newton, who laid the foundations of modern physics and astronomy.

The industrial revolution catalyzed the trajectory of modernity, replete with factories and railways, fundamentally altering production and lifestyle. This transformative era began in Britain in the late eighteenth century and spread across the globe, bringing about unprecedented changes in technology, society, and the economy. The steam engine, invented by James Watt, revolutionized transportation and industry, enabling the mass production of goods and the development of urban centers. The invention of the telegraph and the telephone further shrank the world, making communication faster and more efficient.

In the digital era, the Internet accelerates the dissemination of knowledge, tightly interweaving the fabric of our global community. The information age has democratized access to information, enabling people from different corners of the world to connect, share ideas, and collaborate on a scale never before possible. Social media platforms, online education, and digital communication tools have become integral parts of daily life, shaping the way we work, learn, and interact with one another.

Yet, the path to sustainable development for human civilization is fraught with challenges. Ecologically, global warming is escalating, glaciers are receding at an alarming rate, and sea levels are creeping upwards. The consequences of climate change are becoming increasingly evident, with more frequent and severe weather events, such as hurricanes, droughts, and wildfires. Projections suggest that some low-lying island nations may vanish beneath the waves by century's end, displacing vast populations and leading to a new class of climate refugees. The loss of these nations would not only be a humanitarian tragedy but also a cultural and ecological catastrophe.

Deforestation is rampant, the Amazon rainforest is contracting precipitously, biodiversity is under threat, and many endangered species are on the brink of extinction, disrupting the ecological equilibrium. The Amazon, often referred to as the "lungs of the Earth," plays a crucial role in regulating the global climate and supporting an unparalleled diversity of life. The destruction of this vital ecosystem through logging, mining, and agriculture has dire implications for the planet's health and the well-being of future generations.

Economically, trade protectionism is on the rise, with certain developed nations erecting trade barriers, impacting industries such as steel and automobiles, and straining the global supply chain. The imposition of tariffs and quotas has led to trade disputes and retaliatory measures, threatening the stability of international trade and economic growth. The divide between wealth and poverty is widening; the world's wealthiest 1% control nearly half of its assets, while billions of people live in poverty, lacking access to basic necessities such as clean water, healthcare, and education. This growing inequality perpetuates social unrest and hampers efforts to address global challenges, such as climate change and public health crises.

Chapter 12
Carry Forward the Common Values of all Mankind Together

The common values of all mankind include peace, development, fairness, justice, democracy, and freedom. These values are not just abstract concepts; they are the bedrock of human civilization, the consensus reached by mankind in the long course of history, and are deeply rooted in everyone's heart. They are the guiding principles that help societies progress and individuals thrive.

Peace is the most fundamental of these values. Looking back at the two world wars, the flames of war consumed millions of lives, entire cities were reduced to rubble, economies collapsed, and the very fabric of civilization was torn apart. The horrors of those conflicts have left an indelible mark on human history. After the war, all countries, realizing the catastrophic consequences of conflict, jointly established the United Nations to maintain world peace and stability. Since then, although local conflicts have occurred from time to time, the overall peace situation has been maintained, and many countries have resolved their disputes through diplomatic negotiations, such as the Six-Party Talks on the Korean Peninsula nuclear issue, which aimed to find a peaceful resolution to the nuclear crisis.

Development is the key to solving global problems. It is the engine that drives progress and improves the human condition. The Asian Tigers—Singapore, Hong Kong, Taiwan, and South Korea—achieved economic take-off through an export-oriented strategy, becoming role models for other developing nations. Over the past 40 years of reform and opening up in China, hundreds of millions of people have been lifted out of poverty, creating a miracle of poverty reduction that has been unprecedented in human history. The economic growth of these countries has not only improved the living standards of their citizens but also contributed to global economic stability and prosperity.

Fairness and justice are the foundation of social harmony. In the United States, the civil rights movement of the 1960s promoted racial equality legislation, such as the Civil Rights Act of 1964, which helped to dismantle legal segregation and discrimination. In India, the struggle of the untouchable class led to the implementation of affirmative action policies, which have gradually improved the lives of millions of

L. Yan, *Value-Based Civilization*, Understanding China,
https://doi.org/10.1007/978-981-96-9490-7_12

people and allowed them to participate more fully in society. Social class mobility is a crucial aspect of fairness, and many countries have seen improvements in this area, with more opportunities for individuals to rise based on merit rather than birthright.

Democracy takes various forms, and each country has developed its own version that suits its cultural and historical context. China promotes people's democracy, where the people participate in all aspects of national governance, from community-level elections to the formulation of national plans, thus fully ensuring that the people are the masters of the country. In India, democracy is characterized by vibrant elections and a robust parliamentary system. In the United States, democracy is marked by a strong tradition of civic engagement and a robust system of checks and balances.

Freedom is a value that is demonstrated within the legal framework. It is the right of individuals to live their lives as they choose, provided they do not infringe upon the rights of others. The rise of Internet social media has allowed the public to express their views on many issues more freely than ever before, and this has led to increased transparency and accountability in governments and corporations. Social media platforms have become powerful tools for social movements, enabling people to organize and advocate for change, as seen in the Arab Spring and the MeToo movement.

But now, there are many obstacles to these common values. Globalization has brought unprecedented economic opportunities but has also led to inequality and cultural tensions. Climate change poses a threat to peace and development, as environmental degradation can lead to resource scarcity and conflict. Technological advancements have raised concerns about privacy and the ethical use of data. Political polarization and misinformation threaten the very fabric of democracy and freedom. Despite these challenges, the pursuit of these common values remains a shared goal for humanity, and it is through cooperation and dialogue that we can overcome these obstacles and continue to move forward.

Chapter 13
Build the Symbiosis and Common Prosperity of Human Civilization

Human civilization is like a rich and colorful garden, where each country and national culture is a unique flower with its own characteristics, blooming together in a harmonious tapestry. Asia, with its ancient and enduring Chinese civilization, takes the Confucian thought of "benevolence" as the core, a philosophy that has guided countless generations. This civilization has cultivated a rich tradition in calligraphy, painting, and poetry that spans thousands of years, each art form containing profound cultural deposits that reflect the wisdom and aesthetics of its people. The Indian civilization, on the other hand, has bred the Buddhist philosophy, which advocates for inner peace and enlightenment. Yoga, a cultural gem of India, has become popular worldwide, promoting physical and mental harmony and well-being.

In the Middle East, the grandeur of mosque architecture stands as a testament to the region's architectural prowess. The intricate designs and towering minarets are not only places of worship but also symbols of cultural identity. The exquisite handicrafts, such as the famous Persian carpets, showcase the region's artistic talent and attention to detail. Moreover, the accurate astronomical calendar developed by ancient scholars in the Middle East demonstrates the unique wisdom and scientific achievements of this civilization.

European civilization, with its roots in the speculation of democracy, the rule of law, and philosophy in ancient Greece and Rome, has evolved through the Renaissance and the Enlightenment, achieving remarkable advancements in science, literature, and art. The influence of European thinkers like Isaac Newton, whose work on mechanics reshaped our understanding of the physical world, and William Shakespeare, whose plays continue to captivate audiences with their universal themes, are testaments to the enduring impact of European culture. The artistic genius of Leonardo da Vinci, whose masterpieces like the Mona Lisa and The Last Supper are celebrated globally, further highlights the richness of European civilization.

Despite the hardships faced by African civilizations due to colonialism, the continent's cultural heritage remains vibrant and full of life. The passionate music and dance of Africa, such as the energetic rhythms of Afrobeat and the traditional dances

L. Yan, *Value-Based Civilization*, Understanding China,
https://doi.org/10.1007/978-981-96-9490-7_13

of various tribes, are expressions of joy and resilience. The exquisite wood carving skills found in regions like Nigeria and Kenya are not only artistic achievements but also a means of storytelling and preserving history. Ancient tribal traditions, including the intricate oral histories and vibrant festivals, continue to be celebrated, showcasing the tenacious vitality of African cultures.

In reality, however, conflicts between civilizations occur from time to time, casting a shadow over the harmonious coexistence of mankind. Differences in religious belief can sometimes cause unrest in various regions, as seen in the Middle East where different sects have differing interpretations of holy land, leading to conflicts and social disorder. Cultural misunderstandings can lead to stereotypes and prejudices, as often depicted in Western film and television works that distort Oriental culture, shaping Asian women as docile and weak images, which can hinder cross-cultural exchanges and understanding.

Behind the veil of economic interests, some developed countries have attempted to assimilate or marginalize weaker civilizations. Hollywood films, for instance, often propagate American values and cultural norms while squeezing the cultural space of others. This cultural imperialism can lead to a homogenization of global culture, where unique local traditions and identities are overshadowed by dominant global narratives. It is crucial for the international community to recognize and respect the diversity of human civilization, fostering dialogue and understanding to ensure that the rich tapestry of global culture continues to flourish in all its colorful variety.

Chapter 14
Create a Common Future of Human Civilization

To create the common future of human civilization is the common expectation and responsibility of all countries and nations in the world. This vision requires a concerted effort to address the myriad challenges that arise from rapid technological advancements. Under the wave of scientific and technological development, cutting-edge technologies such as artificial intelligence and quantum computing are changing the world at an unprecedented speed, revolutionizing industries, and transforming daily life. However, these advancements also bring many new challenges, such as the employment structure adjustment caused by artificial intelligence, where traditional jobs are being replaced by automated systems, leading to a need for workforce retraining and education. Privacy protection problems have also emerged, as personal data is increasingly collected and analyzed, raising concerns about surveillance and data misuse. Furthermore, the impact of quantum computing on the existing encryption system poses a significant threat to cybersecurity, potentially rendering current encryption methods obsolete and exposing sensitive information to new vulnerabilities.

In the face of these challenges, countries need to work together to formulate unified science and technology ethics, establish international standards, and create regulatory frameworks. This collaborative approach will guide science and technology to be good, ensuring that innovation serves humanity and does not become a hidden danger. It is essential to foster an environment where technology is harnessed for the betterment of society, promoting sustainable development and equitable access to the benefits of technological progress.

In the field of education, efforts should be made to cultivate a new generation with a global vision and cross-cultural communication skills. To achieve this, educational systems around the world must evolve to include international perspectives and foster a deeper understanding of diverse cultures. International schools and joint education projects can be further promoted, offering students a curriculum that includes world culture comparison, global research, and encourages participation in international

L. Yan, *Value-Based Civilization*, Understanding China,
https://doi.org/10.1007/978-981-96-9490-7_14

exchange activities. Programs such as simulating the United Nations or international science competitions can provide students with hands-on experience in global collaboration, making them aware of the interconnectedness of our world and the importance of working together to solve global issues. These experiences will help students set up a human destiny community consciousness and understand that they shoulder the mission of creating a better future for all.

Economic cooperation is a top priority for fostering a common future. Countries should strengthen trade relationships, reduce tariff barriers, and promote the deep integration of industrial chains to realize complementary advantages. This can be achieved through bilateral and multilateral trade agreements that facilitate the free flow of goods and services. Emerging economies and developing countries can use the Belt and Road Initiative and other platforms to strengthen infrastructure connectivity, enhancing their own development capacity and joining hands with more established economies to achieve common prosperity. Investment in infrastructure, such as transportation networks, energy systems, and digital communications, will not only improve the quality of life for citizens but also create new opportunities for economic growth and innovation.

In terms of cultural exchange, in addition to educational initiatives, countries can promote mutual understanding and respect through various cultural activities. Festivals, art exhibitions, music concerts, and film screenings that celebrate the diversity of world cultures can be organized on an international scale. These events provide a platform for artists and creators to share their work and for people to experience the richness of different traditions. Cultural exchange programs can also facilitate the movement of artists, scholars, and cultural professionals, allowing them to collaborate and learn from one another. By promoting cultural diplomacy, countries can build bridges of friendship and cooperation that transcend political and economic differences, fostering a more harmonious global community.

References

1. Yan Lijin, Value Civilization: Digital Technology and the Community of Shared Human Destiny (Chinese Edition), Electronics Industry Press, 2023.
2. Karl Marx, Friedrich Engels, The German Ideology, translated by Central Compilation and Translation Bureau, People's Publishing House, 2019.
3. Friedrich Engels, Dialectics of Nature, People's Publishing House, 2018.
4. Karl Marx, Friedrich Engels, The German Ideology, translated by Central Compilation and Translation Bureau, People's Publishing House, 2019.
5. Zhang Jun, The Ontological Foundation of Value Philosophy, People's Publishing House, 2018.
6. Hamilton, Human Civilization, translated by Zhang Junfeng, Petroleum Industry Press, 2015.
7. Max Scheler, Formalism and Material Value Ethics in Ethics, translated by Ni Liangkang, Life, Reading, and New Knowledge Sanlian Bookstore, 2004.
8. Brian Arthur, The Nature of Technology, translated by Cao Dongming, Wang Jian, Zhejiang People's Publishing House, 2014.
9. Lewis Mumford, Technics and Civilization, translated by Chen Yunming, Wang Keren, Li Huashan, China Architecture & Building Press, 2009.
10. John von Neumann, Computer and the Human Brain, translated by Wang Wenhao, Commercial Press, 2022.
11. Ray Kurzweil, The Singularity Is Near, translated by Li Qingcheng, Dong Zhenhua, Tian Yuan, Mechanical Industry Press, 2022.
12. Zhao Guodong, Yi Huanhuan, Xu Yuanzhong, Metaverse, Zhongyi Press, 2021.
13. Kim Sangyun, The Age of Metaverse, CITIC Publishing Group, 2022.
14. Ray Kurzweil, The Singularity Is Near, translated by Li Qingcheng, Dong Zhenhua, Tian Yuan, Mechanical Industry Press, 2022.
15. Karl Jaspers, The Origin and Goal of History, translated by Li Xuetao, East China Normal University Press, 2018.
16. T.S. Ashton, The Industrial Revolution: 1760–1830, translated by Li Guanjie, Shanghai People's Publishing House, 2020.
17. Erwin Schrödinger, What Is Life, translated by Zhang Butian, Commercial Press, 2018.
18. Jean Baudrillard, Symbolic Exchange and Death, Yilin Publishing House, 2006.
19. Central Compilation and Translation Bureau, Marx-Engels-Lenin-Stalin Works: Volume 2, People's Publishing House, 2009.
20. Yan Xiaofeng, Creating a New Form of Human Civilization, Social Sciences Academic Press, 2022.
21. William Petty, Selected Works of Petty's Economics, translated by Chen Dongye, Commercial Press, 1981.
22. Max Weber, The Protestant Ethic and the Spirit of Capitalism, translated by Yan Kewen, Shanghai People's Publishing House, 2018.

L. Yan, *Value-Based Civilization*, Understanding China,
https://doi.org/10.1007/978-981-96-9490-7

23. Zhang Shuguang, Modern Changes in Social Order, People's Forum: Academic Frontiers, 2015, Issue 7.
24. Thomas Ashton, The Industrial Revolution: 1760–1830, translated by Li Guanjie, Shanghai People's Publishing House, 2020.
25. Arnold P. White, Bai Fulian, Technology in World Civilization, translated by Zhu Tongyue, CITIC Publishing Group, 2023.
26. Lewis Henry Morgan, Ancient Society, translated by Yang Dongchun, Zhang Liyuan, Feng Hanjie, Commercial Press, 1971.
27. Lewis Mumford, Technics and Civilization, translated by Chen Yunming, Wang Keren, Li Huashan, China Architecture & Building Press, 2009.
28. William Bernstein, The Great Trade, translated by Hao Nan, CITIC Publishing Group, 2020.
29. Hamilton, Human Civilization: A Humanist Enlightenment, translated by Zhang Junfeng, Petroleum Industry Press, 2015.
30. Peter Singer, Animal Liberation, CITIC Publishing Group, 2018.
31. Zhang Xiaoyu, Technology and Civilization: Our Time and Future, Guangxi Normal University Press, 2021.
32. Stefan Zweig, The World of Yesterday, translated by Pan Zili, Gao Zhongfu, Times Literature and Art Publishing House, 2018.
33. World Openness Report 2022, China Social Sciences Press.
34. China Academy of Information and Communications Technology, Global Digital Economy White Paper (2022), 2022.
35. Paul Kennedy, The Rise and Fall of the Great Powers, translated by Jiang Baoying et al., China Economic Publishing House, 1989.
36. Klaus Schwab, The Fourth Industrial Revolution: The Forces of Transformation, 2016, CITIC Publishing Group.
37. Kevin Kelly, Out of Control: The Future of Humanity, translated by Zhang Xingzhou, Chen Xinwu, Wang Qin, et al., Electronics Industry Press, 2018.
38. Norbert Wiener, Cybernetics, translated by Wang Wenhao, Commercial Press, 2020.
39. Alvin Toffler, The Third Wave, CITIC Publishing Group, 2018.
40. Jiang Qiping, Outline of New Civilization Theory (Volume 1), Commercial Press, 2012.
41. Klaus Schwab, The Fourth Industrial Revolution: The Forces of Transformation, CITIC Publishing Group, 2016.
42. Cai Jian, Value Internet: Economic Transformation Beyond Blockchain, Tsinghua University Press, 2021.
43. Jean Baudrillard, Symbolic Exchange and Death, translated by Che Jinshan, Yilin Publishing House, 2006.
44. Robert Triffin, Gold and the Dollar Crisis: The Future of Convertibility, translated by Chen Shanglin, Lei Da, Commercial Press, 1997.
45. Xing Jie, Zhao Guodong, Xu Yuanzhong, et al., Metaverse Tokens: The Passport to the Future, Zhongyi Press, 2021.
46. Yuval Noah Harari, Homo Deus: A Brief History of Tomorrow, translated by Lin Junhong, CITIC Publishing Group, 2017.
47. Rebecca Harding, Jack Harding, The Great Power Trade Game: Win-Win Strategies in the Digital Age, translated by Yu Dongmei, Contemporary World Publishing House, 2021.
48. Jean Baudrillard, Symbolic Exchange and Death, translated by Che Jinshan, Yilin Publishing House, 2006.

MIX
Papier aus verantwortungsvollen Quellen
Paper from responsible sources
FSC® C105338

If you have any concerns about our products,
you can contact us on
ProductSafety@springernature.com

In case Publisher is established outside the EU,
the EU authorized representative is:
Springer Nature Customer Service Center GmbH
Europaplatz 3, 69115 Heidelberg, Germany

Printed by Libri Plureos GmbH
in Hamburg, Germany